To Carlos,
hoping that he will have
a wonderful time with AREVA!
May 2008

Almost all climate scientists agree that by mid century the intolerably hot summer of 2003 when over 20,000 died in Europe will be the norm. They expressed their recognition of the severity of the near future climate in the 2007 Intergovernmental Panel on Climate Change report and noted that we are the cause of it.

Our present civilisation is wholly dependent on a constant and reliable supply of electricity. Few seem to realise that major cities such as London, Paris or New York would die within a week without it. To continue to enjoy our civilised lives we must maintain the flow of electricity but we face the dilemma that our main source of energy is fossil fuel combustion, the source of our discomfort.

Romantic notions of clean renewable energy are beguiling but wholly impractical as are attempts to sequester a significant part of the carbon dioxide emitted by combustion. We have one well tried and safe source of energy, nuclear. I welcome and applaud the book by Bertrand Barre and Pierre-Rene Bauquis, which shows us that we need not fear it and that it is the safest and most environmentally friendly source of that vital product, electricity.

James Lovelock

Publications of

BERTRAND BARRE
Tout sur l'énergie nucléaire (AREVA)

PIERRE-RENE BAUQUIS
& EMMANUELLE BAUQUIS
Petrol and natural gas
Understanding the future
(Ed. Hirle)
Pétrole & gaz naturel,
comprendre l'avenir
(Ed. Hirle)
Russian & Spanish versions

Graphic design
Studio Mark C. Ashcroft

ISBN 978-2-914729-53-6

EDITIONS HIRLÉ
131, rue Boecklin
67 000 Strasbourg
Tél. +33 3 88 41 85 01
9, rue Saint Yves
75 014 Paris

www.editions-hirle.com
editionhirle@wanadoo.fr

BERTRAND BARRE PIERRE-RENE BAUQUIS

UNDERSTANDING THE FUTURE NUCLEAR POWER

PREFACE
ANNE LAUVERGEON

EDITIONS HIRLE

Preface

To borrow an expression from Nadine Gordimer, Nobel Prize in Literature 1991 from South Africa, "dread attends the unknown." This legitimate reaction has harmed nuclear power in the past, for too long a matter left to the specialists. There are reasons for this: the defense origins of the research that led to its development carried the seeds of a culture of secrecy.

More than that, the architects of nuclear programs undoubtedly spent more time doing things right than talking about it!

But in a democracy, nothing lasting can be built without the consent of a large majority.

Today, energy has moved to the forefront on the political scene and in the media: there are concerns about continuity of access to fossil fuels and how their costs will evolve; security of supply continues to be an imperative; reducing greenhouse gas emissions a must.

Obviously, there are no miracle cures for these issues, but rather a set of interconnected solutions.

Nuclear power is one of them, as it generates large amounts of domestic energy reliably and without CO_2.

It should be one of the pillars of energy for the future, just like improved energy efficiency or the increased use of renewable energies.

This is a complex issue, difficult to render in non-technical terms, and often perceived as a taboo subject or a source of controversy.

Just as for climate change, energy should be the subject of in-depth "political" debate, in the best sense of the word.

The authors of this book work hard to lay the groundwork for such debate by covering every aspect of nuclear power. They don't hide their opinions, nor do they dodge controversial issues. The quality of the many illustrations keeps the discussion simple yet not simplistic. All these virtues empower this book to promote the adoption of a new energy culture, now more necessary than ever to tackle the major challenges of the 21st century.

Anne Lauvergeon
Chief Executive Officer of AREVA

Contents

1. What is Energy? 9
2. What is Nuclear Energy? 15
3. How Nuclear Energy was Discovered 25
4. Military Applications of Nuclear Energy 33
5. The Production of Electricity from Nuclear Power 55
6. The PWR Fuel Cycle 63
7. Nuclear Power in the World 83
8. Nuclear Energy: the Economic Aspects 101
9. Other Applications of Nuclear Energy 115
10. Nuclear Power and the Environment 130
11. Safety, Radiation Protection and Waste Management 133
12. Future Nuclear Systems 151
13. Prospects for Nuclear Fusion 159
14. A World Vision of Energy: 2000 – 2100 171
Conclusion 186

What is Energy?

The Different Forms of Energy 11
The Different Energy Sources 12
Primary Sources and Energy Carriers 12
World Energy Balance 13
The Decline of Oil 14

Energy is the property of a physical system that is capable of producing work or an action

We all have a basic idea of what the word energy means. When we feel "full of energy" it means we are full of the power we need to spring into action, relatively speaking! Scientists use the word in practically the same way. Energy is the property of a physical system that is capable of producing work or an action. This could be the production of mechanical force (energy to get a car or a train moving) or of heat (energy for cooking a meal or heating up the bath water).
The amount of energy produced depends on the intensity of the work or the "power" and the time over which the power is supplied.

Energy = Power x Time

Example 1

Take a car with an 80 horsepower engine consuming 6 litres of fuel when driving along at 100 kph. It will therefore consume 6 litres in an hour, 12 litres in 2 hours, 18 litres in 3 hours and so on. Its power remains constant (80 hp) but the energy it uses up is proportional to how long it is used.

Example 2

When we put on a electric light such as a big 1000 watt (1 kilowatt) halogen lamp, it will have consumed 1 kWh (1 kilowatt hour) by the end of the first hour, 2 kWh after two hours and so on. This is why electricity companies bill their customers in kilowatt hours, in other words the energy they have used over the month or a quarter.

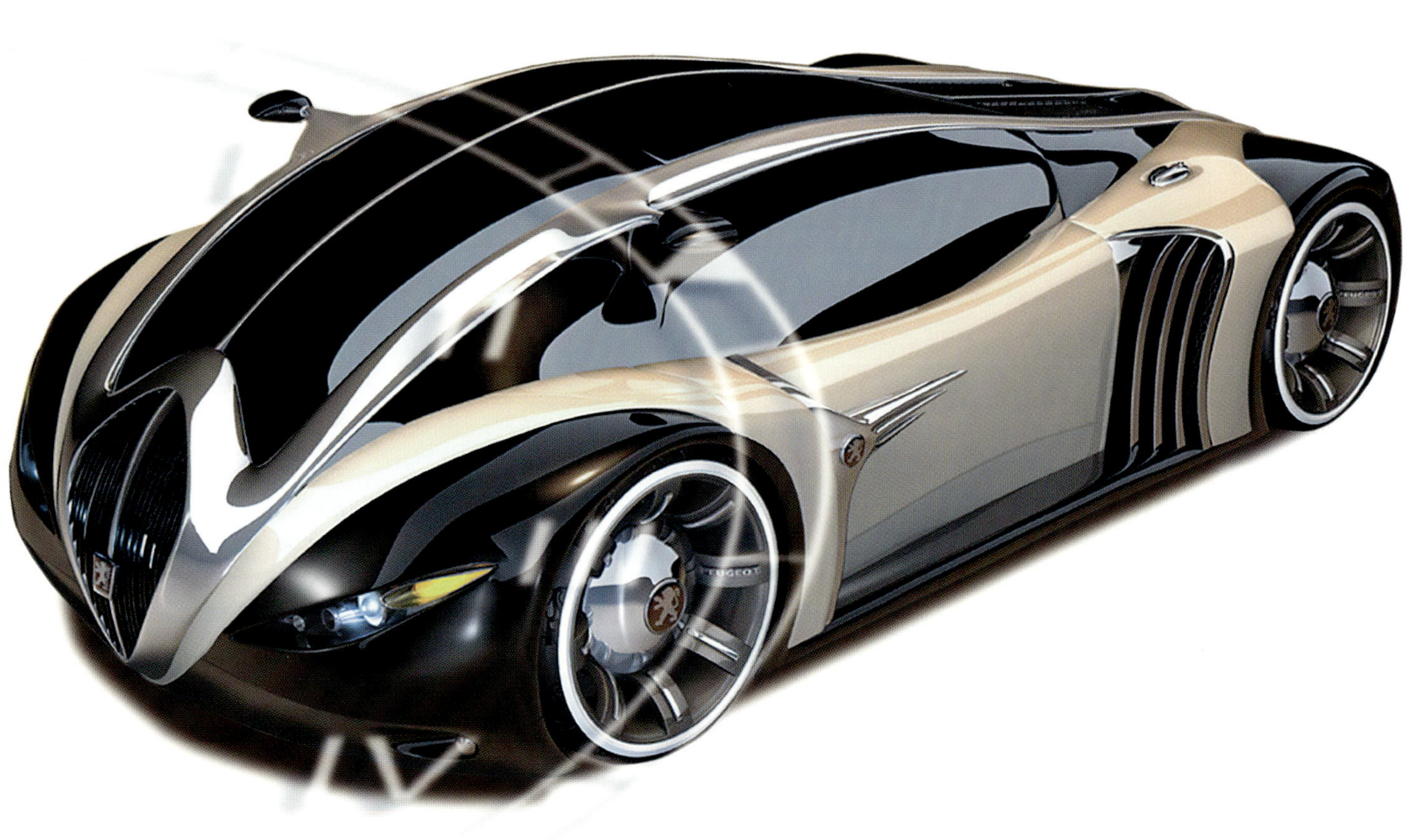

The Different Forms of Energy

Energy exists in a number of different forms:

Mechanical energy, such as in muscles and motors.

Heat energy, such as in central heating.

Energy in the form of waves, such as in light and lasers.

Kinetic energy, such as in a golf ball or a javelin.

Chemical energy such as in food and fuels.

Nuclear energy, such as in nuclear power plants.

These are all in fact different forms of the same thing and this is why they can all be measured using the same units. Anything that man or nature does uses these different energies, which are continually changing from one form into another.

Power Units and Equivalencies

Power	Equivalencies
Kilowatt (kW)	*1 kW = 1.35 hp*
Horsepower (hp)	*1 hp = 736 W = 0.736 kW*
Examples	
Electric lamp – 100 W	*0.1 kW*
Electric iron – 1000 W	*1 kW*
Small car – 60 hp	*44 kW*
Large car – 200 hp	*147 kW*
1 large windmill – 1 MW	*1,350 hp*
1 nuclear plant – 1000 MW	*1,350, 000 hp*

Energy Units and Equivalencies

Energy	Equivalencies
Kilowatt hour (kWh)	*1 kWh = 1.35 CVh*
Metric ton of oil equivalent (toe)	*1 toe = 11.632 kWh*
Barrel of oil equivalent (boe)	*1 toe = 7.3 boe*
Joule	*1 toe = 41.8. 10^9joule*
1 kWh	*100g of gasoline or diesel fuel*
1 toe	*1,000 m^3 of natural gas*

An example: man

A man at rest consumes around 2.5 kWh per day, about the same as a 100 watt electric lamp (100W x 24 hrs = 2.4 kWh). Energy consumption increases five and even tenfold during intense sport.

.

The different Forms of Energy

Energy is produced from a variety of primary sources, which have to be processed to provide energy in a form we can easily use. The forms of energy we can buy are known as commercial energy.

Primary and original sources of energy should not be put in the same category as the so-called energy carriers, which are by-products of the former.

These energy carriers are much easier to use and transport than the primary sources.

Forms of Energy Sources

Primary Sources:

Fossil Energies: *concentrated, non-renewable energies (solar energy stored over geological time)*

- Coal
- Petroleum
- Natural gas

Nuclear energies: *highly concentrated energies*

- Fission : current nuclear power plants
- Fusion : sun, future nuclear power plants ?

Renewable energies: *diluted or diffuse renewable energy forms, (apart from geothermy, they come from the solar energy that reaches the Earth's surface).*

- Hydropower
- Solar • Wind • Biomass • Geothermal

Energy Carriers:

"Secondary" energy produced from primary sources

- Electricity
- Hydrogen

The World's Energy Balance

Every country can total up all the energy sources it uses and work out its individual energy balance. To allow everyone to understand and compare the different energy balances, they have to be measured using the same unit. Up to the 1950s the tce (ton of coal equivalent) was used, before it was replaced by the toe (ton of oil equivalent). The pic chart at the bottom of the page gives an energy balance for the world in 2005.

It should be pointed out that such balances are arbitrary to a certain extent, and therefore not accurate, as they rely on the conventions used to convert the other energy sources into oil equivalent (see chapter 14).

The choice of the toe as the universal unit bears witness to the fact that oil was the main driver of the technical and social revolution in the 20th century. The future of oil in the 21st century will be hampered by two major problems:

- The level of reserves (the "peak oil" problem).
- The problem of greenhouse-effect gas emissions and climate change.

World oil production will fall off long before 2050. We are going to have to develop massive new sources to replace it and be sure these sources do not produce greenhouse-effect gases

Before coal came into daily use, Europe made use of energy from wood, water-mills and windmills. The picture shows a well-known French windmill that inspired Alphonse Daudet to write "Letters from my Windmill". A typical windmill produced about 1 hp, the equivalent of about 2% of the power of a small car today. (Renault Twingo, Fiat Lupo, Volkswagen Polo, Peugeot 107).

The Decline of Oil

All specialists agree that world oil production will decline long before 2050 and maybe as soon as 2010-2020. It would then be necessary to massively develop new energy sources to replace it which do not emit greenhouse effect gases. Those energies will be, on the one hand, renewable energies (wind, solar, biomass) and on the other hand, nuclear energy.

These two forms of energy are not in competition, but are actually complementary.

However first of all we need to explain what nuclear energy is.

The World's Energy Sources

(commercial sources, 2005)

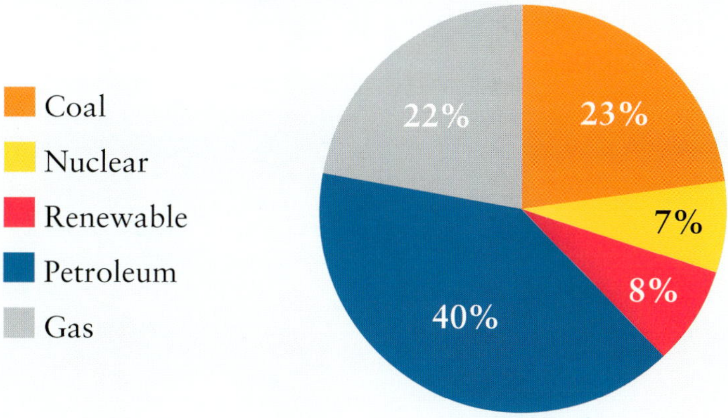

E=MC²

What is Nuclear Energy?

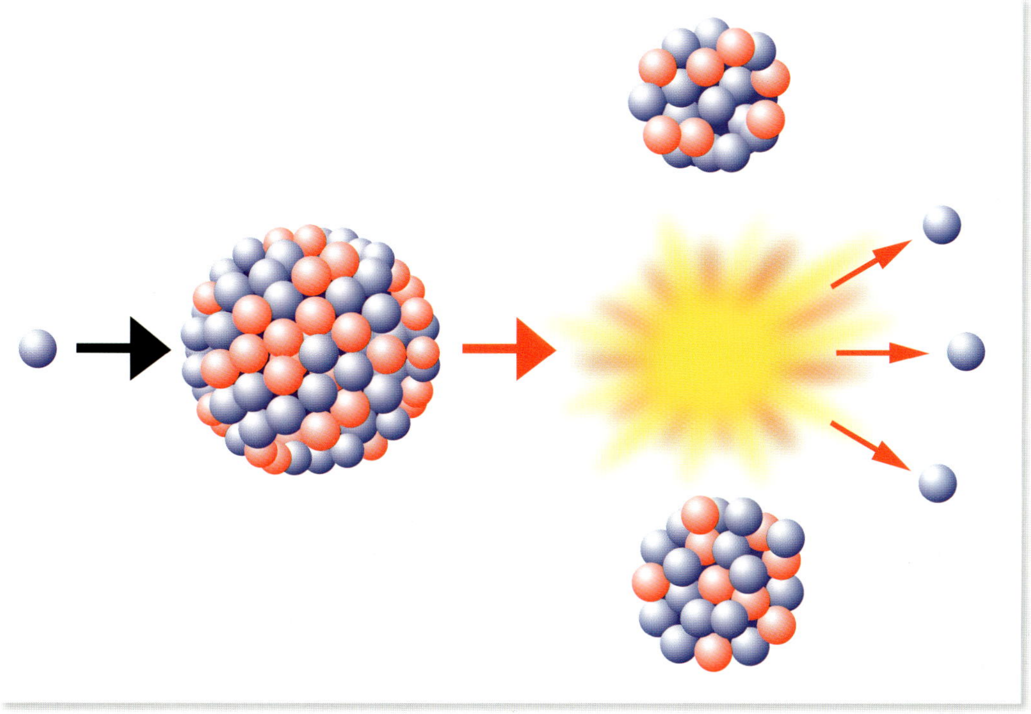

Atoms, Molecules, Nuclei, Isotopes 16

Radioactivity 17

Natural Radioactivity, Artificial Radioactivity 18

Units: Becquerel, Gray and Sievert 20

Origin of Nuclear Energy 21

Fission 22

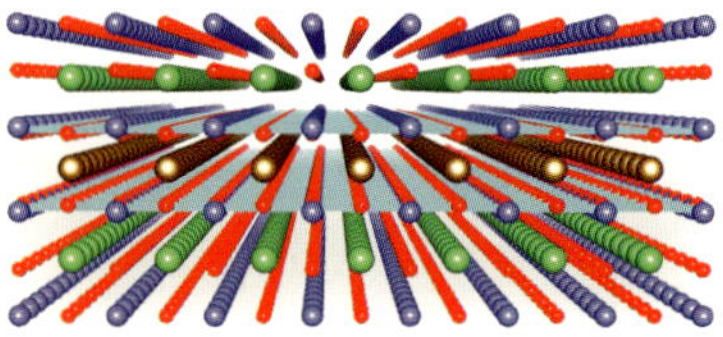

Atoms, Molecules, Nuclei, Isotopes

Ordinary matter around us, whether solid, liquid or gaseous, is made up of atoms. Virtually the whole of an atom's mass is concentrated in the nucleus, which consists of two particles, neutrons and protons, collectively known as nucleons. Protons have a positive charge. Orbiting around the nucleus is a cloud of electrons, negatively charged particles of tiny mass. The electrical charges of the electrons offset those of the protons, so the atom is electrically neutral

Even if you have actually studied physics, it is very difficult to realize that matter as we know it is made up mainly of empty space, with very little true matter (the protons and neutrons, etc.). A useful analogy is to see the space occupied by matter within a substance (be it, for example, oil or a metal teaspoon) as the equivalent of a fly (the particles) buzzing around within St. Paul's Cathedral (the empty space).

The chemical properties of an atom are determined by the number of protons in its nucleus. This number can range from 1 for hydrogen to 92 for uranium. Two atoms that have the same number of protons, but a different number of neutrons are known as isotopes of an element. For instance, uranium has two main isotopes: uranium 238, ^{238}U, with 92 protons and 146 neutrons, and uranium 235, ^{235}U, which has 92 protons and 143 neutrons.

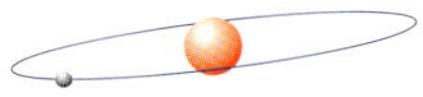

Hydrogen atom

Helium atom

Uranium 238 atom

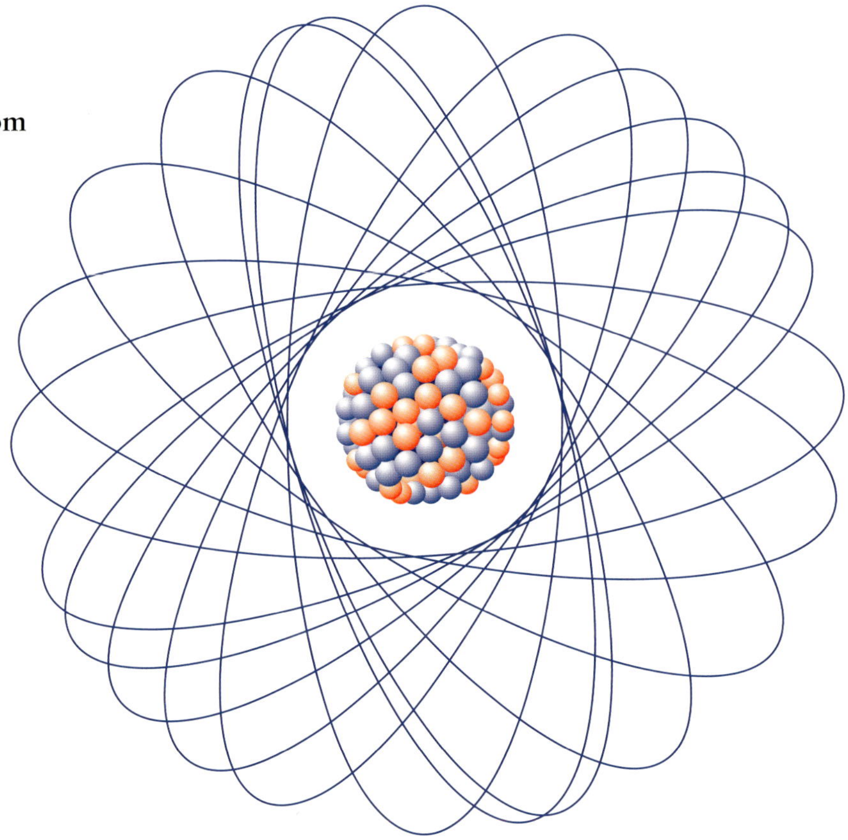

Atoms can share part of their electron cloud to form molecules. Chemical energy which only involves electrons, is produced when bonds between atoms or molecules are broken or formed. Nuclear energy on the other hand, is released during the splitting or fusing of atomic nuclei through fusion, radioactivity and fission.

Some atoms, or rather their nuclei, can split or fuse in processes which involve changes in the nucleons in the nuclei and this is the origin of the two forms of nuclear energy: fission and fusion. Energy released in nuclear reactions is about 1 million times higher than in chemical reactions, which do not involve the nucleons.

Nuclear energies are roughly 1 million times larger than chemical energy, which does not involve nucleons

Radioactivity

Some atomic nuclei are unstable - they spontaneously transmute themselves into other nuclei, with different properties. When this transformation occurs, the nucleus emits radiation, in the form of particles or electromagnetic rays. Such nuclei are said to be radioactive. The radioactivity of any given nucleus is specific and characterized by the type of radiation emitted, its energy, and the speed at which the nucleus decays.

While it is impossible to predict the exact moment at which any single radioactive nucleus will undergo decay, we know with great statistical precision the amount of time, T1/2 it takes for half of the nuclei to decay. This time is called the half-life of the radioactive nucleus.

The main types of radiation emitted by radioactive nuclei are the following: alpha (α) radiation is the high-energy emission of an helium nucleus, made up of two protons and two neutrons. α radioactive nuclei are mostly heavy nuclei, and often have long or very long half-lives. For instance, uranium 238, ^{238}U, is α radioactive and its half-life is 4.5 billion years, about the age of our solar system (as old as the Sun and its planets)!

Beta (β) radiation is the simultaneous emission of one electron and one neutrino, a small particle of tiny mass, the interaction of which with matter is so low as to be insignificant. The kinetic energy of electrons can vary considerably, and they are called "hard" or "soft" beta rays. The half-life of a β radioactive nucleus is very variable, but often rather short, around a few years.

A neutrino can travel right through the whole of our globe without even deviating from its trajectory.

Gamma (γ) radiation is the emission of photons by excited nuclei which can thus release their excess energy and move down to a less excited state. A gamma ray is a form of electromagnetic radiation like visible light or X-rays, with higher energy.

α and $\beta + \gamma$ radioactivity

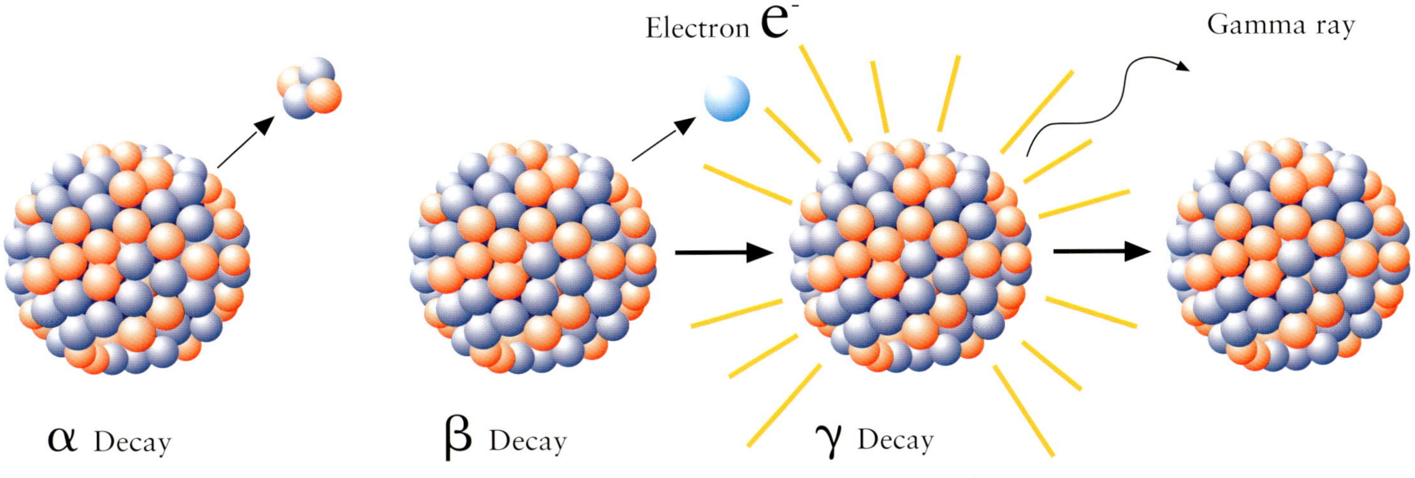

There are other types of radiation, but less common.

Natural and Artificial Radioactivity

Henri Becquerel.

Radioactivity was discovered accidentally in 1896 by Frenchmen Henri Becquerel, when he noticed that the uranium salts he was working with emitted invisible rays that were able to pass through protective black paper and leave an impression on photographic plates.

The term "radioactivity" was coined by Pierre and Marie Curie, when they discovered that uranium ore contained two other elements, polonium and radium, which were much more radioactive than uranium. All three were jointly awarded the Nobel Prize in 1903.

In 1934, Irène Curie, Marie's daughter, and her husband, Frédéric Joliot, bombarded a thin aluminum plate with α rays from a polonium source and obtained phosphorus 30, the first artificial radioactive element. They were award the 1935 Nobel Prize for their work. Other researchers soon followed in their footsteps, bombarding all the nuclei they could lay their hands on with various forms of radiation and managing to produce a multitude of artficial radioactive nuclei.

Radioactive atoms, both "artificial" or "natural", emit the same kind of radiation, with the same effect on living organisms. Protection in either case is the same: do not go too close, restrict exposure time and use a protective shield when necessary.

Radioactive atoms, both "artificial" or "natural", emit the same kind of radiation, with the same effect on living organisms

Pierre Curie.

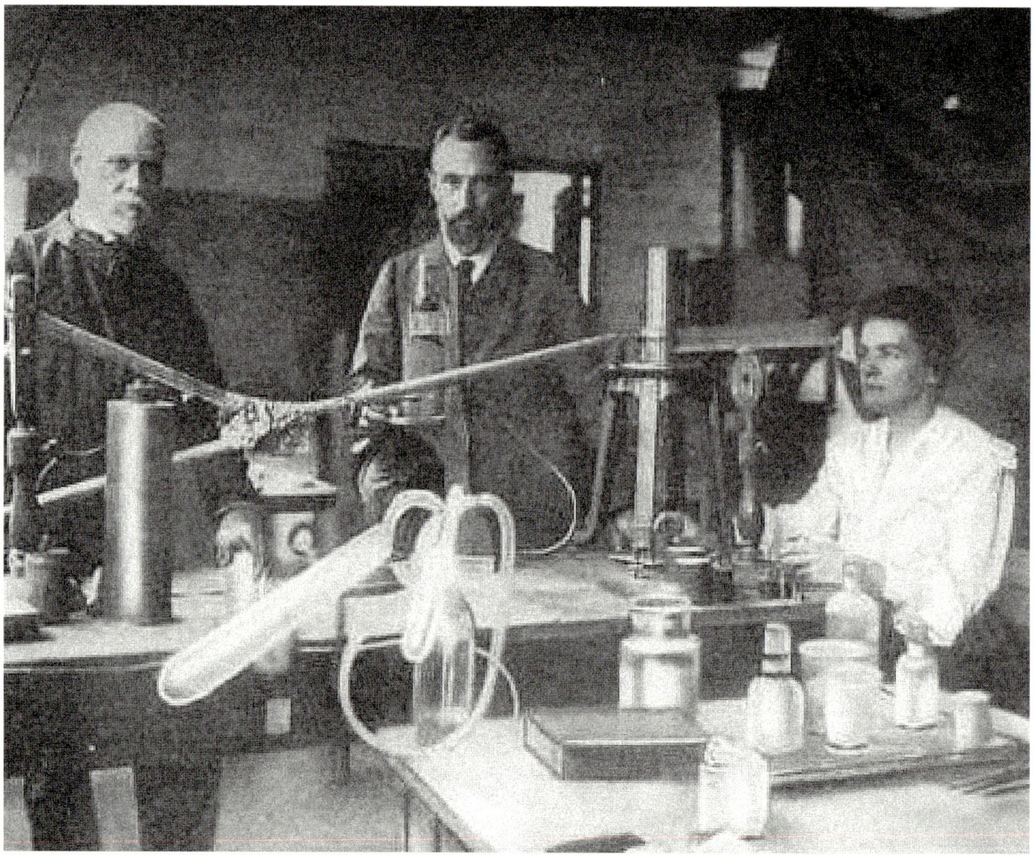

Pierre and Marie Curie in their laboratory.

Irène and Frédéric Joliot-Curie in the middle, with, from left to right, Guéron, Perrin, Stohr, Dautry, Kowarski and Goldschmidt.

I am radioactive, and so are you!

Three types of radioactive elements occur naturally :

- Radioactive elements dating back to the begining of the Earth with a half-life so long that they have not yet fully decayed. These include uranium 235, uranium 238, thorium 232 and potassium 40;
- Decay products of the above (which is why we talk about uranium and thorium families) like radium and polonium;
- Radioactive elements continuously produced by reactions between cosmic rays and elements on the Earth's surface or in the atmosphere. These include tritium (hydrogen isotope described in chapter 13) and carbon 14[1], the concentration of which is used to date organisms by measuring the proportion of the isotope remaining. Carbon 14 is produced from a reaction with nitrogen in the high atmosphere.

Terrestrial potassium therefore contains a natural radioactive isotope, as does atmospheric carbon. As our skeleton absorbs potassium and our tissues contain carbon, all of us are radioactive. Radioactivity in an average adult is close to 8000 Bq, roughly half from potassium 40 and half from carbon 14. That does not mean however we are being turned into radioactive waste!

[1] *A small amount of plutonium is being continuously produced, so it is not just artificial*

Radioactivity in an average adult is close to 8000 Bq, roughly half from potassium 40 and half from carbon 14. That does not mean we are being turned into radioactive waste!

Radiation Units : Becquerel, Gray and Sievert

The activity of a substance is the number of its atoms which undergo decay per second and it is measured in becquerels. 1 Bq = 1 decay per second (irrespective of the type of radiation). One becquerel is a tiny unit.

The absorption of one joule of radiation energy by one kilogramme of matter is expressed in grays.

1 Gy = 1 joule/kilogram.

Radiation energy will have different biological effects according to the type of radiation and the type of organ it is absorbed by.

The biological effects of radiation absorbed by living tissue are measured in sieverts, Sv, or more commonly in millisieverts, mSv, and the result is called the equivalent dose. For instance, the equivalent dose received by an average European from natural sources is about 3 mSv per annum.

The equivalent dose received by an average European from natural sources is about 3 mSv per annum

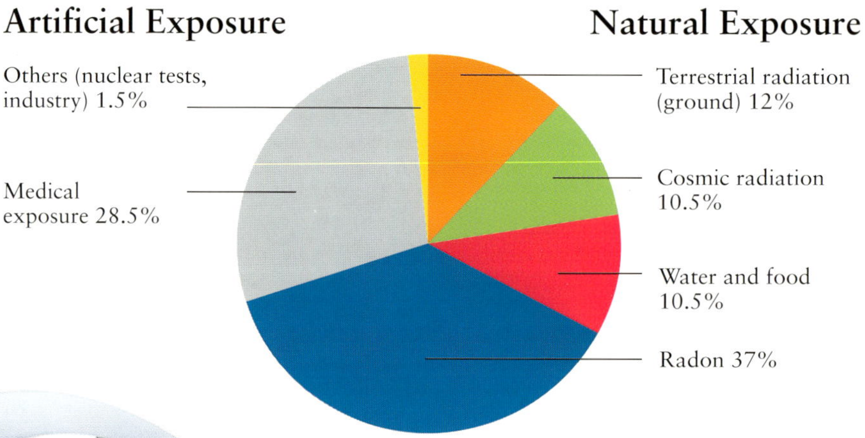

Did you know?

Apart from natural radiation, the main source of radiation is medical treatment and imagery. In developed countries, it accounts on average for 28.5% of total exposure and practically all exposure excluding natural radioactivity.

Origin of Nuclear Energy

Everybody knows Einstein's formula:

$E = mc^2$, where c^2 is the square of the speed of light in a vacuum, a huge number.

The formula shows that when a body has a mass, it has a massive amount of energy associated with this mass and that a slight variation in mass represents an enormous variation in energy.

The total mass of a nucleus is thus lower than all the masses of all its protons and neutrons put together and the difference in mass is the binding energy of the nucleus, generally expressed in millions of electron volts or MeV.

The most stable combination of nucleons happens to be a medium-sized nucleus. This means that if two light nuclei can be fused to form a medium-sized nucleus, some binding energy will be released, which can be recovered and if a heavy nucleus can be split into two medium-sized nuclei, the resulting energy can also be harnessed. This is what is known as nuclear fusion and nuclear fission.

Binding energy is considerable and, as we already saw, is about a million times higher than the energy involved in chemical reactions between atoms and molecules.

This is why the fission of a gram of uranium or plutonium produces more energy than the combustion of a ton of oil, gas or coal (14 tonnes = 1,000 kilograms = 1 million grams)

The fission of one gram of uranium or plutonium produces more energy than the combustion of one ton of oil

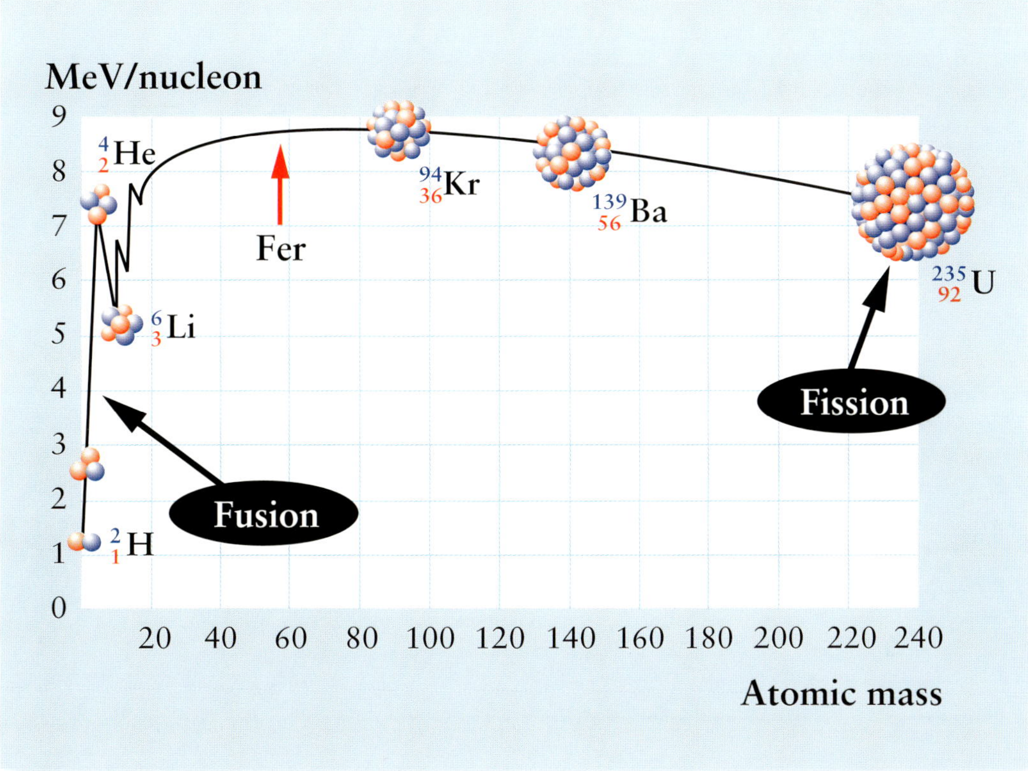

Nuclear Fission

Fission of a uranium 235 nucleus

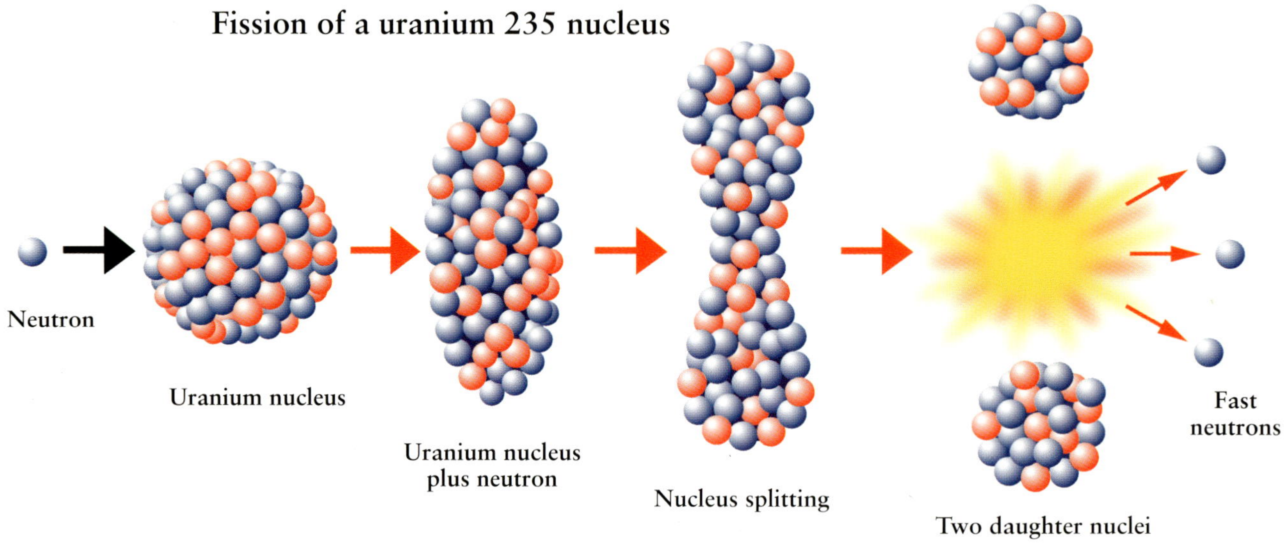

Some heavy nuclei are fissile: after absorbing a neutron, the composite nucleus is too excited to get rid of its extra energy through simple decay and like an oversize drop of water, it violently splits into two unequal fragments, while at the same time ejecting two or three neutrons. This phenomenon is called fission.

The total mass of the fragments and the neutrons is slightly smaller than the mass of the composite nucleus and energy is therefore released. Most of the energy is released as kinetic energy of the fragments. When the fragments slow down, this kinetic energy becomes heat and it is this heat which is transformed into electricity in nuclear plants.

Neutrons ejected during one fission reaction can, in turn, be absorbed by other neighboring fissile nuclei, thus creating a chain reaction. If this chain reaction is left to develop exponentially, it leads to a nuclear explosion.

When the reaction is controlled and stabilized, it produces an energy source and this is what happens in a nuclear reactor (see chapter 5).

Chain Reaction

1. A neutron is about to hit the nucleus of a uranium atom.

2. The uranium nucleus splits (fissions) into several smaller atoms, releasing heat and several more neutrons.

3. The chain reaction begins: the neutrons hit other nuclei, causing them to fission, and so on.

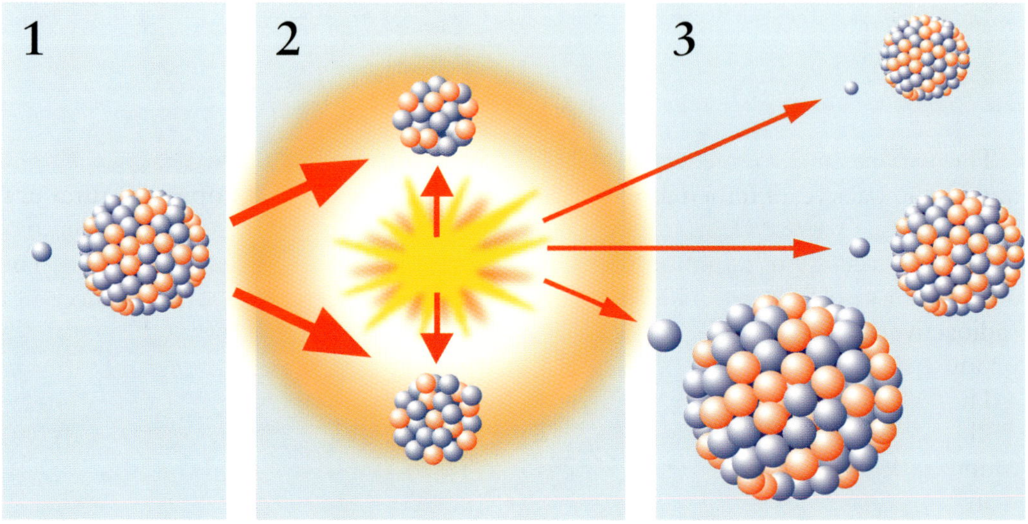

The two fission fragments are almost always radioactive, and therefore undergo successive radioactive decays. The fission fragments and their daughter elements are called fission products. Most high-level radioactive waste is made up of fission products (see chapter 11).

The only readily fissile nucleus found in the Earth's crust is an isotope of uranium, with an atomic mass of 235. This isotope constitutes 0.7% of natural uranium. On the other hand, when a nucleus of the other isotope of natural uranium, uranium 238, absorbs a neutron, it transmutes into fissile plutonium. This plutonium can also generate energy in a nuclear reactor. When another naturally occurring isotope, thorium 232, absorbs a neutron it also turns into a fissile isotope, uranium 233. Uranium 238 and thorium 232 are said to be fertile materials.

Machining a compoment of a nuclear reactor.

Pierre and Marie Curie

How Nuclear Energy was discovered

The fifth Solvay Conference in Brussels, October 1927, from left to right, back row to front row 1. A. Piccard -2. E. Henriot - 3. P. Ehrenfest - 4. E. Herzen - 5. T. De Donder - 6. E. Schrödinger -7. E. Verschaffelt - 8. W. Pauli - 9. W. Heisenberg - 10. R.H. Fowler - 11. L. Brillouin - 12. P. Debye - 13. M. Knudsen - 14. W.L. Bragg - 15. H.A. Kramers - 16. P.A.M. Dirac - 17. A.H. Compton - 18. L.V. de Broglie - 19. M. Born - 20. N. Bohr - 21. I. Langmuir - 22. M. Planck - 23. M. Curie - 24. H.A. Lorentz - 25. A. Einstein - 26. P. Langevin - 27. C. E. Guye - 28. C.T.R. Wilson - 29. O.W. Richardson

The Atom and its Components. 26
Fission 28
The Oklo Phenomenon 30

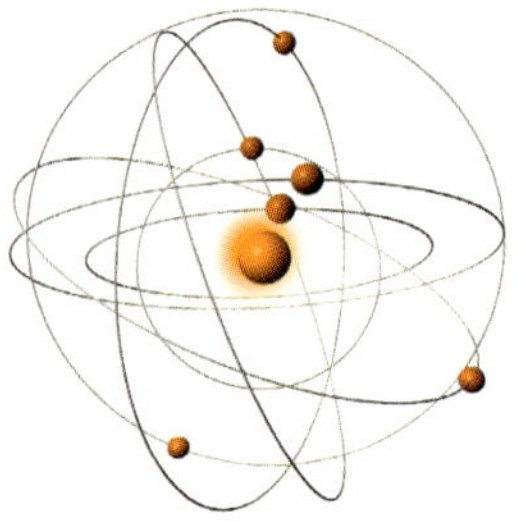

In 1910 the scientific world adopted the Rutherford model, which represents the atom as a miniature solar system

The Atom and its Components

Matter in its three normal states - solid, liquid and gaseous, is made up of atoms, which can bind together to form molecules of varying complexity. Although this is a notion that seems perfectly obvious to us now, it took a long time for it to be accepted. The idea actually dates back to the ancient Greeks, in particular Democritus and Epicurius, who thought that all matter was made up of invisible, indivisible elements called *atomos* (which means indivisible in Greek). It took, however, another couple of thousand years before 19th century scientists (Gay-Lussace, Berzelius and especially Dalton) were able to confirm the theory.

It was not until 1910 that Rutherford produced his model, which represented the atom like a kind of mini solar system, with a central, positively charged nucleus that contained practically the whole mass of the atom, surrounded by a cloud of tiny negative-charged particles called electrons. The sum of the electrical charges of the electrons offset the charge of the protons, meaning the atom is neutral. The chemical type and physical properties of the atom are determined by the positive charge of its nucleus, i.e. the number of protons, and this number is represented by the letter Z. Hydrogen has an atomic number Z =1, Helium 2 and Uranium 92. The notion had already been predicted in 1896 when Mendeleyev came up with his periodic table classifying the elements to illustrate recurring chemical properties.

Another significant breakthrough was made by at the turn of the 19th century when Henri Becquerel and Marie and Pierre Curie discovered radioactivity, the property certain atoms have of emitting radiation. In 1905, Einstein published his theory of relativity, with the famous formula showing the equivalence between energy and mass (see chapter 2) $E = mc^2$.

Ernest Rutherford

Henri Becquerel

The Greek philosopher Democritius

Niels Bohr

James Chadwick

Radium bomb Chicago 1952.

Rutherford's work was taken a step further by Danish physicist Niels Bohr, who in 1913 published a paper showing that the nucleus is not a single particle, but one made up of a number A of nucleons, each of almost identical mass.

A number Z of these nucleons consists of positively charged protons, while the complement (A-Z) is made up of electrically neutral neutrons. The neutron was discovered by James Chadwick in 1932 (whence his nickname Jimmy Neutron). The nuclei of two isotopes therefore contain the same number of protons, but a different number of neutrons, which does not actually much affect their physico-chemical properties.

In 1933, Frédéric and Irène Joliot-Curie discovered artificial radioactivity, and everyone wanted to be part of the game,

In 1933, Frédéric and Irène Joliot-Curie discovered artificial radioactivity and a lot of scientists started bombarding various targets to produce artificial elements

Otto Hahn

with scientists bombarding all kinds of targets to produce artificial elements. The neutron was considered an ideal projectile as it is electrically neutral and therefore not repulsed by either the electrons or the protons of the nucleus.

Over in Italy, in 1934, Enrico Fermi discovered that if neutrons were slowed down by collisions with non-absorbing light nuclei (water or paraffin), these slow neutrons could then be more easily captured by other nearby nuclei. This is the principle of moderation.

Fermi then proceeded to systematically bombard elements with a neutron source. When he tried bombarding uranium, he thought he had produced a transuranian element, heavier than uranium. It needed another 4 years of continuous experiments before Otto Hahn and Fritz Strassmann came up with the answer, in 1938. They found that a uranium nucleus, when bombarded with neutrons, did not disintegrate as per classical theory, but instead split into two unequal parts. Hahn and Strassmann had discovered fission. Lise Meitner and Otto Frisch calculated that this fission would release a considerable amount of energy, far larger than first thought.

Fritz Strassman

Lise Meitner

Fission

In spring 1939, Frédéric Joliot's team showed that during fission, free neutrons were emitted and they understood that these free neutrons could trigger a new fission event, which would in turn emit more free neutrons, thus making a chain reaction. They considered two applications were possible: either a hugely powerful explosive produced by an exponential chain reaction or the stable production of large amounts of energy through a controlled chain reaction. In May 1939, Joliot, Hans Halban, Lew Kowarski and Jean Perrin secretly registered three patents describing the two concepts. At the same time, Niels Bohr came up with the hypothesis that only uranium 235, which makes up but 0.7% of natural uranium, could undergo fission. With the outbreak of the Second World War on 1 September 1939, research into nuclear fission moved over to the USA, where it was carried out in conditions of utmost secrecy.

Fermi was awarded the Nobel Prize in 1938, following which he emigrated to New York to get away from Mussolini's fascist regime. Soon after his arrival, he began working at Columbia University, where, with Leo Szilard, he studied the possibility of producing a chain reaction with uranium using very pure graphite as a moderator.

Szilard, a Hungarian refugee, fearing that Nazi Germany was developing the atomic bomb, convinced Einstein to write a letter to Roosevelt in 1939 to warn him of the danger. In June 1941, when it was becoming increasingly difficult for the USA to stay out of the war, Roosevelt gave the go-ahead for the Manhattan project. Fermi moved to the University of Chicago in early 1942 to join the project and in November 1942 construction began in a squash court under the Staggs Field stadium of CP1, Chicago Pile 1, made with graphite bricks with uranium fuel. On December 2, 1942, a coded phone call was made containing the famous message "The Italian navigator has lan-

Enrico Fermi.

1.9 billion years ago, part of what is now the Oklo uranium mine in Gabon underwent spontaneous nuclear fission, producing the same self-sustaining nuclear chain reactions as are now produced in man-made atomic piles.

The Oklo Phenomenon

Before 1972, everyone thought that the first nuclear chain reactions were produced in 1942 in the CP1 reactor. However, scientists in the French Pierrelatte enrichment plant noticed a small discrepancy in the amount of ^{235}U isotope in uranium samples compared to the norm. By tracing the anomaly back through the processing stages, they were astonished to find that there were natural reactors in the Oklo uranium mine, the origin of the samples, that dated back some two billion years. The reactors operated at a relatively low power level and over 150,000 or so years they consumed about 5 tons of ^{235}U. In all, 17 reactors were found in Oklo, but nowhere else in the world. The uranium was then several thousand meters under the surface, in highly porous, water-saturated sandstone, under high pressure, although it has now risen to the surface. The concentration of ^{235}U in the natural uranium back then was about the same as that in enriched uranium, used today in reactor fuel. The actinides, however, have not moved for 2 billion years, which is encouraging for the any future geological storage of nuclear waste (see chapter 11).

ded in the new world...", which signalled the first self-sustained nuclear chain reaction to be achieved by man. As it turned out, though, Nature had already produced a nuclear chain reaction a couple billion years before, in Gabon.

The aim of the Manhattan project was to produce not nuclear energy, but the A-bomb

The aim of the Manhattan project was to produce not nuclear energy, but the A-bomb, as we shall see in chapter 4. The project was run by the military and the first atomic explosion, Trinity, was carried out on July 16, 1945, near Alamogordo in New Mexico.

Shortly after the war ended, moves were made in several countries to set up organisations to foster the development of the new form of energy. In France, de Gaulle created the Atomic Energy Commission, while Britain set up the Atomic Energy Research Establishment, the forerunner of the United Kingdom Atomic Energy Authority, created in 1954.

The McMahon Act in the USA put nuclear technology under the control and management of the civilian-run Atomic Energy Commission and also put a strict ban on the transfer of atomic technology to any other power, even allies. Canada created its Atomic Energy of Canada Limited in 1952.

The USSR set up the Kurchatov Institute of Atomic Energy, under the control of the Russian Ministry of Atomic Energy and 3 years later saw the start-up of the F1 reactor, the design of which was similar to the CP1.

The first reactor built outside the USA, however, was the ZEPP (1945), the work of Canadian and British scientists. The French followed suit 3 years later with the Zoe experimental reactor, which went critical on December 15, 1948. The scientists had done their job, now it was up to the engineers.

Igor Kurchatov

K5 Enrichment plant in Oak Ridge, Tennessee (USA).

French President Vincent Auriol in front of Zoe, the first French experimental reactor.

Plutonium production reactor, Hanford (USA) 1944.

Frédéric Joliot-Curie's press conference where he announced the resuls of the work that led to the construction of the Zoe pile and the new prospects they offered for the future.

Military Applications of Nuclear Energy

World War II 35

The Manhattan Project 37

How Nuclear Weapons Work 40

A-Bombs 40

The Effects of Nuclear Weapons 44

H-Bombs 45

Nuclear Arsenals in the World 46

The Proliferation of Nuclear Weapons. 50

Are there any non-Proliferating Technologies ? 53

Even today, most of us are unable to make up our minds and decide if the use of the atom bomb by the United States was a good thing or a disastrous decision

The one thing about nuclear energy that has perhaps remained indelibly engraved on our conscious and our subconscious is that its first application was military.

The atomic bomb was used for the first and only time in 1945, on Hiroshima (August 6,) and Nagasaki (August 9,).

The bombs caused massive casualties (240,000 deaths), but put an end to the war with Japan. On August 14, 1945, the Japanese government announced the country's unconditional surrender and signed the Instrument of Surrender on September 2, 1945 on board the USS Missouri in Tokyo Bay.

There have been lots of books and discussions about the making and use of the nuclear bomb, but even today, most people cannot make up their mind whether the dropping of the bomb was justified or not. Those in favour claim that the bomb helped cut short the sufferings of the Second World War, while the subsequent "balance of terror" between the USA and the USSR, went a long way to preventing a third world war.

A lot of people think, however, that the use of the bomb was morally reprehensible and opinion was deeply divided even among the scientists working on Manhattan Project.

We will not in this chapter try to reopen the debate, but instead will explain how the first Weapons of Massive Destruction were made, both from the historical and scientific perspectives. We shall then endeavour to give an overview of the development of military applications of nuclear energy since the Second World War.

Signature of the official act of capitulation of Japan, aboard the USS Missouri, on September 2, 1945, in Tokyo Bay.

The Second World War

By an extraordinary coincidence, it was just at the time that the Second World War became inevitable (1938-1940) that nuclear physicians became convinced that it was technically possible to produce extremely powerful nuclear explosions.

The equivalence between energy and matter ($E=mc^2$) had been established by Einstein in 1905, and it meant that a kilogram of matter could in theory produce 25 billion kWh of energy, the equivalent of 10 million tons of a powerful explosive like TNT.

The actual technical aspects of such a process were, however, still the subject of debate in the early 1930s.

It is striking how the Nazi rise to power took place at the same time as scientists were finding the answers to harnessing nuclear energy. Hitler was appointed Chancellor of Germany in January 1933 and the official persecution of the Jewish population began just a few months later, in May. It was in 1934 that Frédéric and Irène Joliot Curie discovered artificial radioactivity (see chapter 3).

The excesses of the Nazi regime caused many German, Austrian, Hungarian and Italian physicists, including Einstein, Max Born, Lise Meitner, Enrico Fermi, Jacob Franck, Eugene Wigner, Leo Silzard and Edward Teller, to leave for England, France or the USA. A handful of nuclear physicists decided to stay in Germany, among them Otto Hahn, Fritz Strassman and Werner Karl Heisenberg, who were working on nuclear fission and were already convinced by 1939 that explosive nuclear reactions could in fact be produced.

Nazi Germany did not give priority to a military nuclear program, preferring to concentrate its resources on the V1 and V2 long-range rocket projects. The decision not to go ahead with a nuclear program was probably taken in July 1942 by Goering, even though a team headed by Werner Karl Heisenberg (Nobel Prize for Physics in 1932) had the scientific knowledge to develop an atomic bomb. It might well have been Philippe Lenard, who also won the Nobel Prize for Physics (1920), who convinced Hitler that nuclear physics was on shaky foundations. If the project had gone ahead, it is worth considering the horrific prospect of what the world might have become had the Nazis had been the first to get their hands on nuclear weapons.

It is worth considering the horrific prospect of what the world might have become had the Nazis had been the first to get their hands on nuclear weapons

Fieldmarshall Gœring with Adolf Hitler.

President Franklin D. Roosevelt.

Albert Einstein
Old Grove Rd.
Peconic, Long Island
August 2nd 1939

F.D. Roosevelt
President of the United States
White House
Washington, D.C.

Sir:

Some recent work by E.Fermi and L. Szilard, which has been communicated to me in manuscript, leads me to expect that the element uranium may be turned into a new and important source of energy in the immediate future. Certain aspects of the situation which has arisen seem to call for watchfulness and, if necessary, quick action on the part of the Administration. I believe therefore that it is my duty to bring to your attention the following facts and recommendations:

In the course of the last four months it has been made probable - through the work of Joliot in France as well as Fermi and Szilard in America - that it may become possible to set up a nuclear chain reaction in a large mass of uranium,by which vast amounts of power and large quantities of new radium-like elements would be generated. Now it appears almost certain that this could be achieved in the immediate future.

This new phenomenon would also lead to the construction of bombs, and it is conceivable - though much less certain - that extremely powerful bombs of a new type may thus be constructed. A single bomb of This type, carried by boat and exploded in a port, might very well destroy the whole port together with some of the surrounding territory. However, such bombs might very well prove to be too heavy for transportation by air.

The United States has only very poor ores of uranium in moderate quantities. There is some good ore in Canada and the former Czechoslovakia. while the most important source of uranium is Belgian Congo.

In view of the situation you may think it desirable to have more permanent contact maintained between the Administration and the group of physicists working on chain reactions in America. One possible way of achieving this might be for you to entrust with this task a person who has your confidence and who could perhaps serve in an inofficial capacity. His task might comprise the following:

a) to approach Government Departments, keep them informed of the further development, and put forward recommendations for Government action, giving particular attention to the problem of securing a supply of uranium ore for the United States;

b) to speed up the experimental work, which is at present being carried on within the limits of the budgets of University laboratories, by providing funds, if such funds be required, through his contacts with private persons who are willing to make contributions for this cause, and perhaps also by obtaining the co-operation of industrial laboratories which have the necessary equipment.

I understand that Germany has actually stopped the sale of uranium from the Czechoslovakian mines which she has taken over. That she should have taken such early action might perhaps be understood on the ground that the son of the German Under-Secretary of State, von Weizsäcker, is attached to the Kaiser-Wilhelm-Institut in Berlin where some of the American work on uranium is now being repeated.

Your very truly

Albert Einstein

General Leslie Groves.

The Manhattan Project

Between 1939 and 1940, leading nuclear physicists (Irène and Frédéric Joliot-Curie in France, Niels Bohr in Denmark, Enrico Fermi, Leo Szilard and others in the USA), became convinced that it was possible to build an atomic bomb.

They decided to warn the government about the danger of Germany being the first to develop such a weapon and they persuaded Einstein to sign a letter on August 2 1939 to Roosevelt.

The letter (see opposite) marked a turning-point and led to the American atomic bomb project, which was given the code-name "Manhattan". The project took place under the greatest secrecy and resulted 6 years later in the bombs that were dropped on Hiroshima and Nagasaki.

Einstein was not involved in the project and only found out about Hiroshima in the newspaper. It was then that he said "if I'd known, I would never have signed the letter."

But it was the surprise attack by the Japanese on December 7, 1941, which destroyed the American fleet in Pearl Harbour, that got America to throw all its resources behind the development of the atomic bomb. In January 1942, Robert Julius Oppenheimer was appointed director of the Manhattan project, despite the misgivings of the FBI, who suspected him of leftist leanings. On top of that, Oppenheimer used to write poetry in Latin and Greek, suspect behaviour for a physicist! In 1953, at the height of McCarthyism, his supposed sympathies for the USSR led him to being stripped of his security clearance and thus his job, the logistics of which were taken over by General Leslie Groves, an outstanding organiser.

J. Robert Oppenheimer.

General Groves in Oak Ridge.

The huge US effort to acquire the A-bomb focused initially on research on the nuclear chain reaction, carried out by four of the leading American universities (Columbia, Princeton, Chicago and Berkeley), work coordinated by Arthur Compton (Nobel prize for physics).

In Chicago, Fermi and Szilard built the first atomic pile (CP1 reactor), which went critical on December 2, 1942, a year after Roosevelt's decision to launch the vast nuclear program.

United States expenditure on the atomic bomb project was enormous

Two Features of the Manhattan Project:

- the sheer size of the project and the speed at which it was carried out,
- the development of two types of A-bomb, which still exist today
- an enriched uranium bomb (Hiroshima)
- a plutonium bomb (Nagasaki).

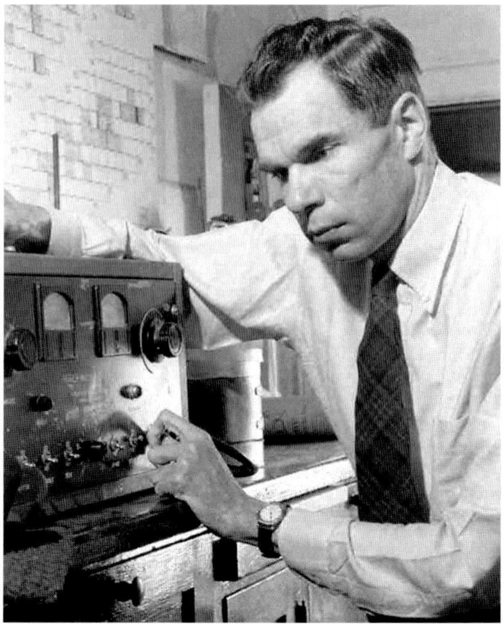

Glenn T. Seaborg and the first Geiger counter.

The first electromagnetic separation plant, using calutrons, in Oak Ridge, Tennessee.

Léo Szilard

The page opposite shows the huge resources used and the remarkable efficiency of the project. It illustrates what people are able to accomplish when they pool their skills and resources.

The War Economy is not a pleasant experience, but it teaches us what we are capable of achieving in the face of adversity.

The Resources used in the Manhattan Project

From left to right : Ernest O. Lawrence, Glenn T Seaborg et J. Robert Oppenheimer.

- about 130,000 people
- total cost put at around 2 billion dollars (about $200 billion in today's terms)
- the construction of the first gaseous diffusion enrichment plant and the first electromagnetic separation plant, featuring a calutron. The plants were used to enrich natural uranium to produce the ^{235}U isotope needed for the gun-type bomb that was to be used on Hiroshima, without being previously tested.
- the construction in 11 months of a graphite moderated reactor at Oak Ridge for producing plutonium (the implosion–type bomb, like the one used on Nagasaki).
- the selection in November 1942, of Los Alamos in New Mexico for assembling the atomic bomb. The site employed some 45,000 workers, followed by about 2,000 civilian scientists and technicians, almost a thousand military personnel and more than 20 Nobel-Prize winners.
- the construction of six plutonium-producing reactors at Hanford, in the State of Washington, between March 1943 and September 1944, under Glenn Seaborg and the US Army Corps of Engineers. Plutonium production started on September 26, 1944.
- Operation Trinity, the first experimental nuclear explosion on July 16 1945, using a plutonium implosion bomb (the same as the Nagasaki bomb). The test was held at Alomogordo in New Mexico, some 350 km from Los Alamos, in front of 150 witnesses, including Oppenheimer, Seaborg, Frisch, Ernest O. Lawrence and Chadwick.
- August 6, 1945, Little Boy was dropped on Hiroshima.
- August 9, 1945, Fat Man was dropped on Nagasaki.
- August 14, 1945, Japan surrendered.
- September 2, 1945, Japan capitulated.

Alamogordo, New Mexico. A steel tower, is erected at ground zero.

Trinity, the first experimental nuclear explosion July 16, 1945, at 5:29:45 a.m.

How Nuclear Weapons work

We saw before that energy from atomic nuclei can be used in two ways:

- fission energy
- fusion energy

Either form can be used to produce extremely powerful weapons:

- atomic bombs or A-bombs which use fission energy,
- thermonuclear bombs or H-bombs which use mostly fusion energy.

While every country has the right to develop nuclear energy for peaceful means, how can they be prevented from going on to make nuclear weapons?

A-Bombs

The diagrams opposite show how the two Manhattan Project bombs worked. The bomb dropped on Hiroshima used virtually pure uranium 235. The design seemed simple enough not to warrant a test before its actual use.

The main explosive for the Nagasaki bomb was plutonium. As the scientists were less sure of the design, they carried out a test, code-named Trinity, on July 16, 1945 at Alamogordo in the New Mexico desert.

Two different plants were required to produce the two bombs:

- an uranium enrichment plant for producing uranium 235 for the first type of A-bomb
- nuclear reactors producing plutonium and a plutonium recovery plant for the second one

Even nowadays, the main concern regarding the proliferation of nuclear explosives is stopping terrorist organisations and rogue states getting hold of the enriched uranium or plutonium which could be subsequently be used for making bombs. The current Iranian crisis (2005-2007) is an illustration of another side of the problem: while every country has the right to develop nuclear energy for peaceful means, how can they prevented from going on to make nuclear weapons? We will be discussing this issue at the end of the chapter.

To put an end to Japanese resistance, President Harry S. Truman decided to drop atomic bombs on two of the country's cities (Hiroshima on August 6, 1945 and Nagasaki on August 9, 1945). Japan capitulated on September 2, 1945.

Little Boy

Uranium Bomb
(Hiroshima,
August 6, 1945)

How Nuclear Bombs work

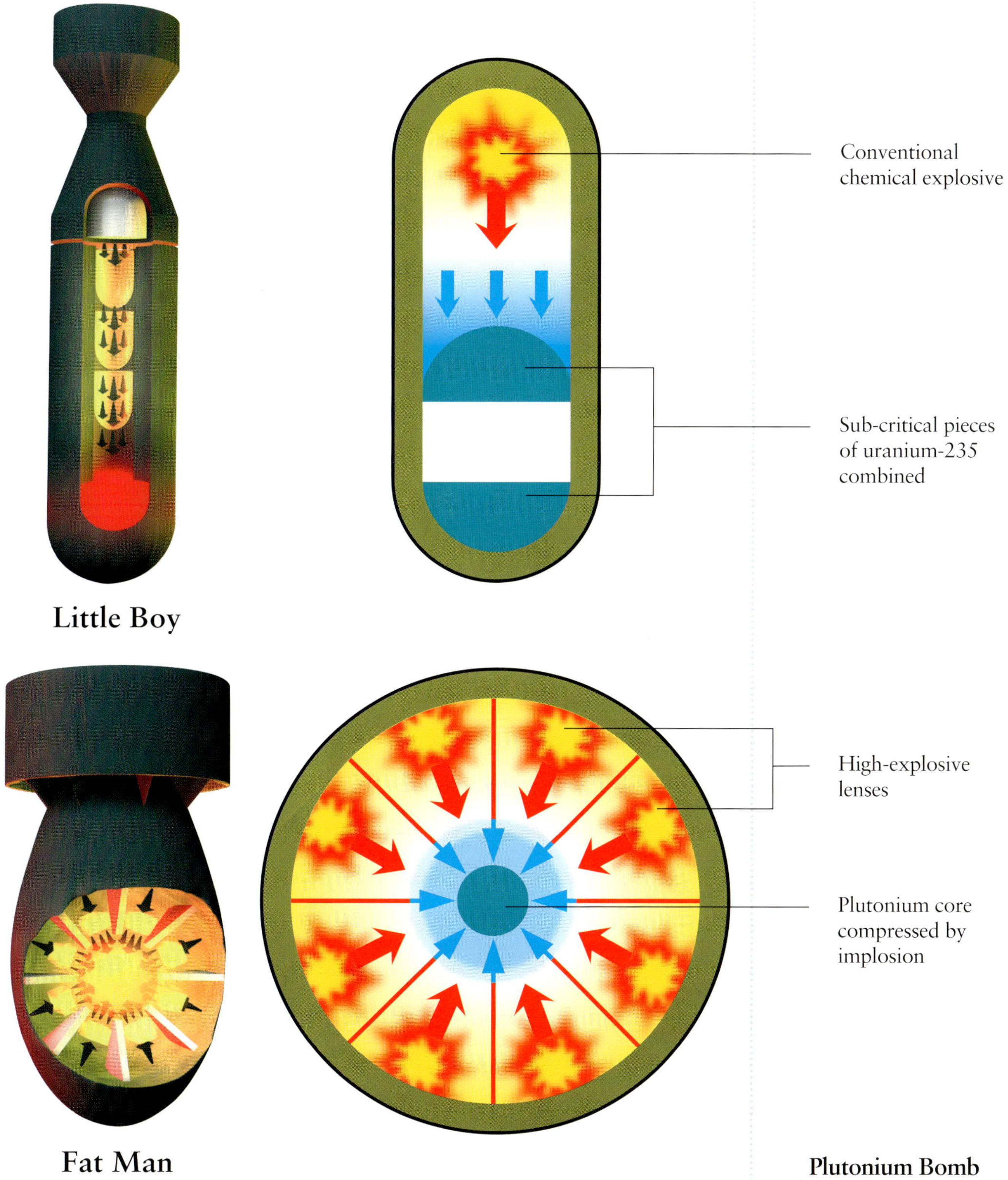

Plutonium Bomb
(Nagasaki,
August 9, 1945)

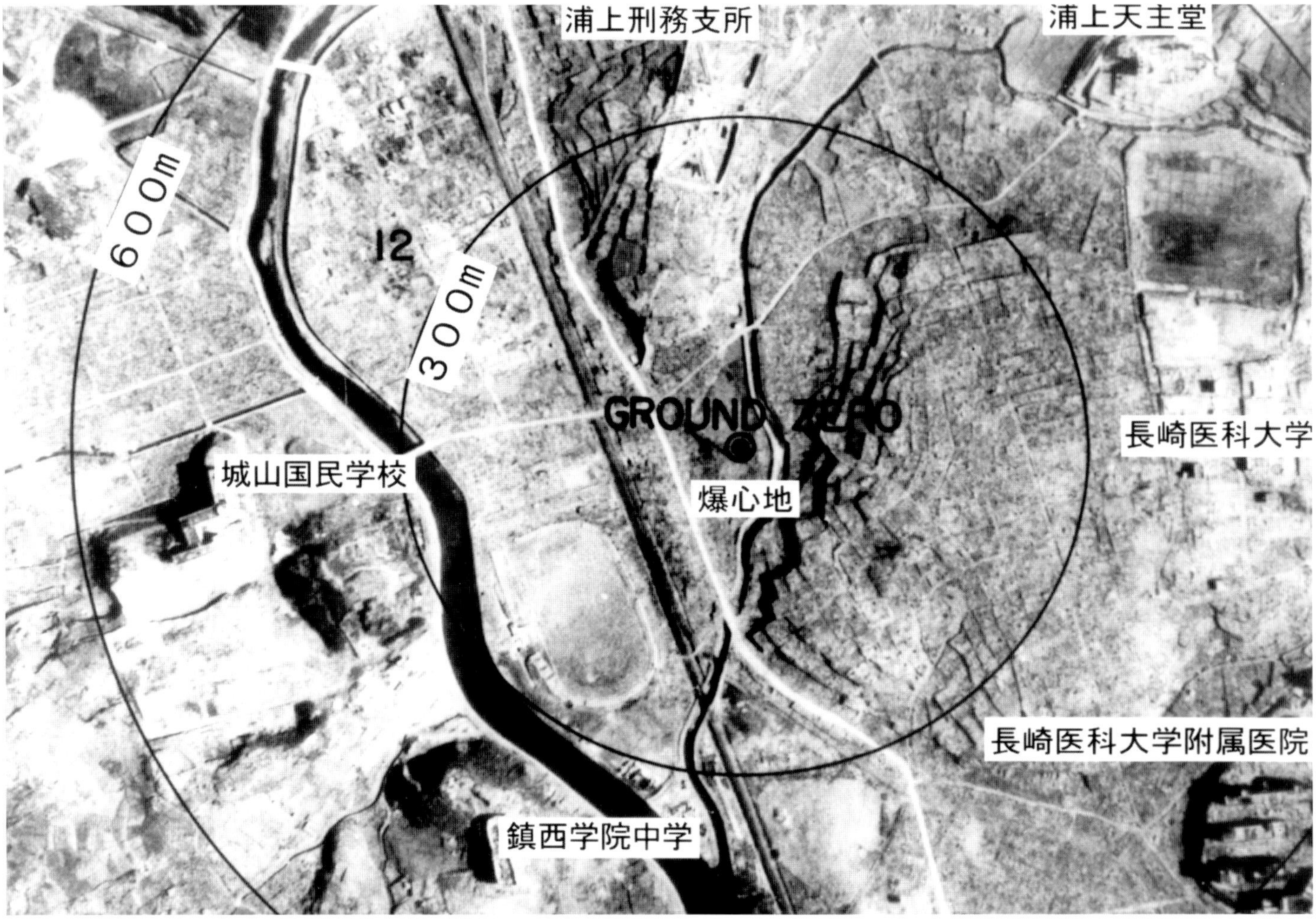

The hypocentre of the Hiroshima bomb (the explosion was at an altitude of 600 meters).

The destructive power of conventional atom bombs comes from the combination of three effects

The Effects of Atomic Bombs

The mushroom cloud is perhaps the classic image of a nuclear explosion. It consists of a rapidly-rising mass (the stem) of low-density gas and debris caused by the initial fireball, which disperses once the gas is no longer less dense than the surrounding air.

The destructive power of conventional atomic bombs comes from the combined effects of the three forms in which the explosive energy is released:

- shockwave and blast effect (about 50% of the energy released)
- thermal effect (about 35% of the energy)
- radiation effect (about 15% of the energy)

The blast and thermal effects can be compared to those produced by conventional explosives dropped on cities. Most of the deaths in Hiroshima and Nagasaki were caused by these two effects. The number of deaths (about 200,000 in all) was about the same as those caused by the carpet bombing of civilian populations using conventional chemical explosives in, for example, Dresden on February 15, 1945, with anything between 40,000 and 130,000 deaths and Tokyo on March 9, 1945, with 70,000 deaths.

The third effect, the radiation due to nuclear weapons, although killing fewer people, is altogether more insidious. Radiation cannot be seen, has no smell and it is often only much later that its harmful consequences begin to be felt.

Shinichi Tetsutani, aged 3 years and 11 months when the atomic bomb exploded.

1,500 meters from ground zero of the Hiroshima explosion.
Shinichi liked going around on his tricycle. On the fateful morning, he was playing outside his house when there was a blinding flash in the sky. Shinichi and his bike were severely burned by the heat of the explosion and the young boy died of his injuries the same night. His father buried him with his bike in the garden of their house, unwilling to abandon his innocent remains to an anonymous grave.
40 years later, in 1985, the remains were transferred to the family vault and the tricycle given to Hiroshima's Peace Memorial Hospital.

The Genbaku dome was the only building left standing in the area where the first atomic bomb exploded. It has been preserved in the same state in the Memorial Park.

Effects of Nuclear Weapons Radiation

Hiroshima and Nagasaki provide the only statistics on nuclear weapons and radiation effects but the results have not been widely publicised.

Thorough studies have been made of the effects of radiation on 120,000 heavily irradiated survivors of the Hiroshima blast, all of whom were within 3 km of the explosion.

The total number of "excess" deaths due to cancer and leukaemia, compared to "natural" deaths from the same causes is less than 500

Assessing the effects of radiation on health, whether in Hiroshima, Nagasaki or Chernobyl, involves measuring excess deaths or effects due to various pathologies (cancers, leukaemia, malformations, cardiac disease) in the irradiated area against a similar, non-irradiated population.

The results are given in the box below.

Excess deaths from different types of cancer and leukaemia amounted to under 500, which is a lot when one takes into account the suffering involved, but which is low compared to the 200,000 deaths caused by the blasts.

It is also low when compared to the public perception of the health consequences of massive radiation by nuclear weapons or nuclear accidents. The facts are rarely given objective or accurate coverage in the press and the exaggerated figures that have now become accepted conventional wisdom make it very difficult to treat the issue of radiation protection (see chapter 11) with the measured objectivity that is needed.

Unfounded claims about nuclear radiation may actually lead to health disasters, such as in the case of Chernobyl, where in all likelihood, some 95% of the deaths were related to psychosomatic problems and only 5% a direct result of radiation.

Hiroshima and Nagasaki did however, leave us a legacy that has had a profound impact on our lives. If we had not been quite so terrified of the thought of nuclear conflict, it is possible that the Cold War could well have led to the unimaginable horrors of a third world war.

But Hiroshima and Nagasaki have also taught us that the risks to our health that nuclear energy poses are quantifiable and limited and that it would be irresponsible to use the fear surrounding it to shy away from using peaceful nuclear energy at a time when we so obviously need it (see chapter 14).

Ginko Biloba

One of the oldest trees in the world, it survived the radiation of the Hiroshima and Nagasaki bombs and symbolises the way nature can adapt to extreme conditions.

The Cumulative Effects, 1946 - 1990, of Radiation on the 120,000 survivors of Hiroshima and Nagasaki

Genetic effects on embryos irradiated "in utero" (3,000):
- very small excess of solid cancers,
- very small excess of leukaemia,
- small excess of mental retardation.

Genetic effects on children born between 1948 and 1954 (77,000)
- no significant teratologic effect (major malformation, cardiac malformations, anencephalies, club feet).

Cancers among the 50,000 who were still alive in 1990:
- less than 10% excess in solid tumours (339 out of 4,687 fatal cancer cases),
- excess leukaemia relatively significant, but low in absolute terms (87 out of 176 fatal leukaemia cases).

Mankind now possesses a cataclysmic power of destruction. Nuclear test at Bikini Atoll (Operation Crossroads) on July 1, 1946.

The week after the Bikini nuclear test, a French designer brought out the world's smallest two-piece bathing costume, the bikini.

H-Bombs

We saw in chapter 2 that there are two forms of nuclear energy: fission energy, used in A-bombs and fusion energy, used in H-bombs.

Bombs that use fusion energy are also called hydrogen bombs, whence the name H-bomb. As no H bomb has fortunately ever yet been used for real, their effects on the population remain unknown. Tests, however, have shown that their destructive power is about 1,000 times higher than A-bombs. In 1961, the USSR tested a bomb, the biggest yet, with an explosive power of around 60 million tons of TNT, compared to the Hiroshima bomb, which was equivalent to about 20,000 tons.

Fusion bombs use the explosion of an A-bomb to heat an adjacent capsule containing a mixture of tritium, deuterium, or lithium deuteride, triggering a fusion reaction.

Nuclear Arsenals in the World

The figure below shows the strong reduction in the nuclear arsenals of the two nuclear superpowers since the collapse of the Soviet Union in 1990.

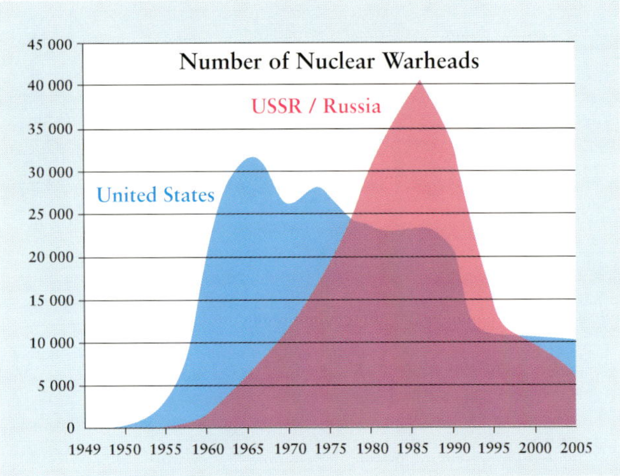

The number of US nuclear warheads peaked at around 30,000 in the early 1960s (compared to about 5,000 for the USSR), while the Soviet Union warhead arsenal topped out at about 40,000 in 1990 (against 25,000 for the USA). There were estimated to be at least 28,000 nuclear weapons in the world in 2005, split as follows:

- USA 10,300, with 4,900 deployed
- Russia 16,000, with 7,400 deployed
- others under 1,400

The other nuclear powers recognised by International Law are the United Kingdom, France and China. India, Pakistan and Israel also possess nuclear weapons, although the latter denies ever assembling any warhead.

Some countries have abandoned their nuclear weapons programs (South Africa, Argentina, Brazil, Libya and Iraq after the first Gulf War), while others, such as North Korea, have been trying to join the nuclear club. Iran claims that its long-covert nuclear program is for peaceful purposes only, although it has not convinced the international community that this is so.

Nuclear arsenals are made up of bombs or warheads and their carriers or launchers. The carriers can be land-based (single or multiple warhead missiles), airborne (bombs or air-to-ground missiles) or aboard ships or submarines.

The submarines and a number of the ships that carry nuclear weapons are themselves nuclear-powered, allowing them to operate submerged and undetected for long periods of time.

While the first nuclear pile went critical in 1942 (CP1 in Chicago), the first American power reactor was the land-based prototype of the one used to power the Nautilus submarine, which would later become world-famous when it completed the first submerged voyage across the North Pole.

Mock-up of a Russian mobile nuclear missile.

US strategic bomber fitted with nuclear weapons.

US H-Bomb.
Atomic Museum
Albuquerque, New Mexico.

Countries possessing Nuclear Weapons

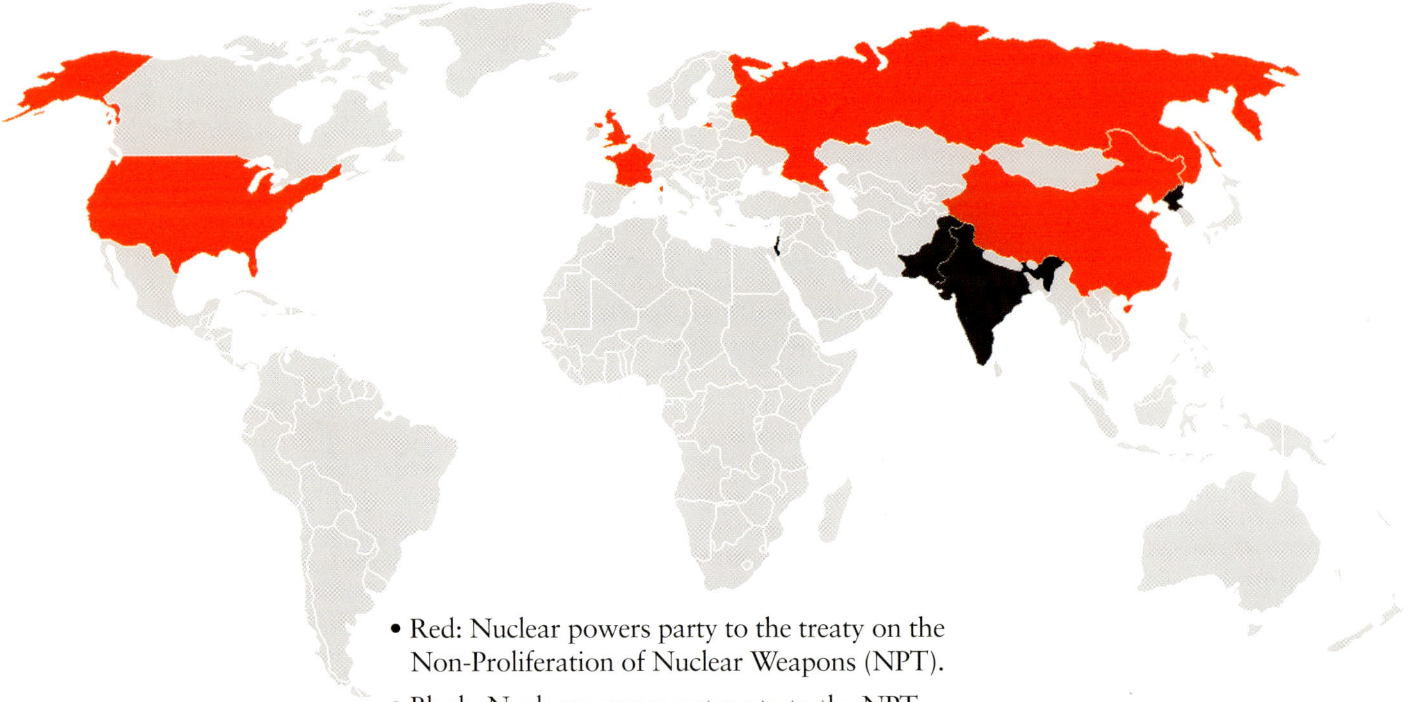

- Red: Nuclear powers party to the treaty on the Non-Proliferation of Nuclear Weapons (NPT).
- Black: Nuclear powers not party to the NPT.

U.S.S. Kennedy in the port of New York.

U.S.S. Nautilus (SSN-571)

The Nautilus was a U.S. Navy attack submarine, the first ever nuclear-powered vessel.

Named after Jules Verne's eponymous submarine, Nautilus' design followed work by Alvin Weinberg. President Truman gave the go-ahead for construction in 1951, under the leadership of then-Captain (and later Admiral) Hyman Rickover. Nautilus was commissioned in 1954 and made her first voyage on January 20, 1955, out of the Groton naval shipyards in Connecticut. The submarine was 91 meters long and had a displacement of over 3,000 tons.

Her nuclear propulsion system allowed her to shatter all submerged speed and distance records and she became the first vessel to complete a submerged voyage across the North Pole.

She was decommissioned in 1980.

In August 1958, Nautilus became the first vessel to complete a submerged voyage across the North Pole

Nuclear powered Warships in operation (2006)

1) Surface ships (aircraft carriers, destroyers 25)

USA :	19 (10 aircraft carriers and 9 destroyers)
USSR/Russia :	5 (1 aircraft carrier and 4 destroyers)
France :	1 (1 aircraft carrier)

2) SSBM submarines :	35
USA :	14
USSR/Russia :	12
UK :	4
France :	4
China :	1

3) Attack submarines :	90
USA :	50
USSR/Russia :	27
UK :	7
France :	6

The Proliferation of Nuclear Weapons in the World

Flag of the IAEA, the International Atomatic Energy Agency.

The term "proliferation" refers to the increase in the number of States in possession of nuclear weapons. It is sometimes used mistakenly to describe the misappropriation of weapons or fissile material by subnational groups. The term "nonproliferation", on the other hand, refers to the political or technical means used to prevent proliferation.

Nonproliferation has had its ups and downs, as summarized below. The United States first tried to protect its military nuclear monopoly by refusing any transfer of civil nuclear technology. When proliferation occurred in the USSR anyway, President Eisenhower changed tack in his famous 1953 "Atoms for Peace" speech and allowed other countries access to reactor technology in exchange for their commitment to using the tech-

Brief History

Year	Country	Proliferation	Nonproliferation
1945	USA	First A Bomb	
1949	USSR	A Bomb	
1952	United Kingdom USA	A Bomb First H Bomb	
1953	USSR USA	H Bomb	 Atoms for Peace
1956	UN		Creation of the IAEA
1957	United Kingdom	H Bomb	
1960	France	A Bomb	
1963	USA/USSR/UK		Moscow Treaty (to limit tests)
1964	China	A Bomb	
1967	China	H Bomb	
1968	France	H Bomb	Non-Proliferation Treaty NPT
1974	India IAEA Exporters	A "Test"	 "Trigger" list London suppliers Group
1990	Irak		Gulf War
1991	South Africa		Dismantling of Weapons, joining NPT
1992	France, China		Join NPT (all NWS Members NPT)
1994	Ukraine, Kazakhstan, Belarussia		Weapons returned to Russia
1995			NPT Extension Initiative
1997	IAEA		Additional Protocol
1998	India Pakistan	H bomb A bomb	
2003	Pakistan, Libya, Iran North Korea	"Bazar" A.Q.Kahn Withdrawal from NPT	Libya joins NPT
2006	Iran North Korea	Ambiguous Program A-Bomb Test	

IAEA headquarters in Vienna, Austria.

nology for civil applications only. During the Cold War, the United Kingdom, then France, and lastly China, joined the Nuclear Weapons States.

In 1968, the Non-Proliferation Treaty, NPT, attempted to freeze the situation by recognizing 5 legitimate nuclear powers but no more. In exchange, the nuclear powers undertook to reduce their arsenal and give free rein to civil technology transfers. The International Atomic Energy Agency, IAEA, set up by the UN in Vienna in 1956, was entrusted with overseeing the peaceful use of nuclear materials. Under IAEA control, technology transfer agreements rose rapidly, each exporter laying down its own conditions for the use of the technologies, facilities, and materials exported.

In 1974, India, which had not signed the NPT, upset the apple-cart by carrying out a "peaceful explosion", using plutonium produced in a heavy water reactor supplied by Canada, the fuel being supplied by the USA. Exporting nations then agreed among themselves to regulate "sensitive" exports. The following period was very positive, in terms of nonproliferation. In 1991, however, a similar shock was felt with the discovery of an extensive clandestine nuclear program in Irak, a country that had signed the NPT. As a result, the powers and inspection carried out by the IAEA were reinforced.

With the end of the Cold War and the collapse of the USSR, the Russian Federation came into possession of the former Soviet Union's military capability. South Africa, Argentina and Brazil (along with the two remaining NWSs, France and China) came into the NPT fold. The 90s as a whole were very positive, but the twenty-first century has got off to the wrong start.

Pakistan, India's rival since partition in 1948, crossed the "nuclear Rubicon" in 1998. North Korea left the NPT at the end of the six month notice stipulated in the Treaty, and resumed plutonium production to carry out a nuclear test in October 2006. When it abandoned its nuclear program to join the NPT in 2003, Libya revealed to the world the existence of a nuclear "black market", led by Abdul Qadeer Khan, the man who, by stealing URENCO's enrichment technology, had become the father of the Pakistani bomb. Apart from Libya, Iran and North Korea had also managed to benefit from allegedly "subnational" transfers of military technology. Iran had to admit that for the last 18 years, it had been developing enrichment technologies in undeclared convert facilities in clear violation of its commitments under the NPT. This led to the 2005-2006 Iranian crisis, during which the other nations asked it to terminate its enrichment.

We should, however, be careful not to throw the baby away with the bathwater. Overall, the NPT has had positive results since its was opened to signature in 1968 and proliferation has been much slower than once feared. Backed by IAEA inspections, the NPT is now the universally-acknowledged cornerstone for all nuclear commerce. Only Israel, India, Pakistan and North Korea have so far failed to come into line. The present system now needs consolidating, but should certainly not be abolished.

Civil Nuclear Industry and Proliferation

The real question that needs to be asked on non-proliferation is whether the development of nuclear energy for civil applications increases or reduces the risks of nuclear weapon proliferation. This question does not deliberately seek to be paradoxical with the words "or reduce". The question really should be asked in those terms and the answer is not necessarily obvious. Except for India, which cleverly played with the clauses imposed on it by Canada, no country opting for proliferation has done so by misappropriating materials or facilities covered by commitments to peaceful use and under IAEA control.

Unfortunately, the first application of nuclear fission was the atom bomb, and not nuclear energy. Fission cannot be

Nobel Peace Prize winner Mohamed El Baradei, director general of the International Atomic Energy Agency (IAEA), waves from the balcony of his hotel after receiving the prize in Oslo, December 10, 2005.

The 3 Pillars of the NPT

Although its structure might appear somewhat unvenly balanced, the NPT, or Nuclear Non-Proliferation Treaty, is a win-win situation for signatory States. In exchange for their undertaking not to acquire nuclear weapons (Art. II) and their acceptance of controls and safeguards (Art. III), Non-Nuclear Weapon States are entitled to benefit from the free transfer of technology for peaceful purposes (Art. IV), while Nuclear-Weapon States undertake to move towards full disarmament, admittedly without any specific timetable. The NPT is therefore all about give and take and the Treaty stands on its three pillars: Nuclear non-proliferation, disarmament and peaceful uses of nuclear technology.

"uninvented", and there will always be a risk of a State or large subnational group deciding to make nuclear weapons and to bear the political consequences, whether or not this State produces nuclear electricity. Neither Israel, Irak or North Korea has ever produced a single nuclear kWh.

Proliferation is 90% political will, and 10% technology

Keeping in mind this unavoidable risk, possessing a civil nuclear facility within its borders may give a State quicker access to the necessary fissile materials. On the other hand, a civil nuclear industry also means international agreements and treaties, commitments not to misappropriate materials, and international inspections with highly sensitive measuring devices which have demonstrated their efficiency: it would be much more difficult to carry out a clandestine program.

Does the expansion of nuclear power increase or reduce the risk of proliferation? As we already saw, the answer is not easy and has to examined case-by-case. The disappearance or a ban on nuclear energy would certainly not remove the risk of proliferation and might actually have the opposite effect. Peaceful nuclear power already helps nuclear disarmament by playing a role in demilitarizing fissile material rendered superfluous by the end of the Cold War: the highly-enriched uranium is converted for civilian use, while military plutonium is recycled for future use as MOX fuel.

Furthermore, once nuclear power becomes a significant part of a country's energy-supply mix, it will have a lot to lose from any ban on international exchanges in the field. You just have to compare the nuclear programs of South Korea and India to see the price the latter has had to pay since it was forced to go it alone in 1974.

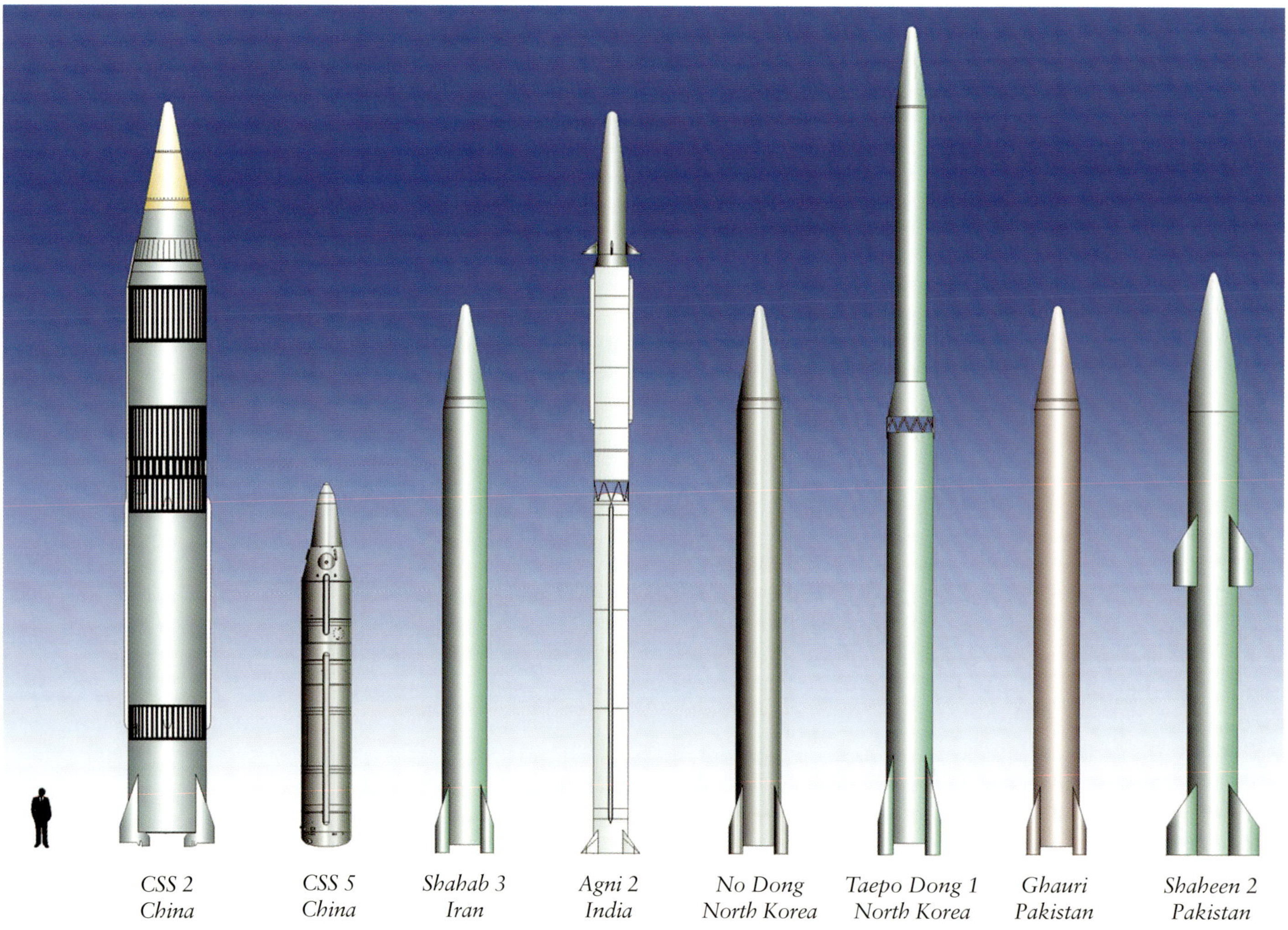

Uranium Enrichment facility in Natanz, Iran. The breaking of the IAEA seals triggered the present crisis (2006).

Are there any "Non-Proliferating" Technologies?

Proliferation is 90% political will, and 10% technology. It is difficult to determine exactly how "proliferating" a technology is. Can an existing facility be put to illegal use or secretly copied? Are we more worried about the manufacture of a single "home-made" device or weapon stockpiling? A lot of factors must be taken into account: ease of access to the pure concentrated fissile material, detectability, throughput, etc.

We can nevertheless hazard two opinions: for a few devices, isotope enrichment is certainly the most discreet method, while for a significant arsenal, a reactor fed with natural uranium, with online fuel loading/unloading is the most effective method.

Conversely, everybody would agree that the PWR and BWR light water reactor technology offers very limited risks of proliferation. In the present system, the sensitive part is the fuel cycle, as shown by the Iranian crisis.

Iran President Mahmoud Ahmadinejad, during his 2006 New Year address on television, asserted Iran's right to continue enriching uranium "for peaceful purposes".

The Production of Electricity from Nuclear Power

Nuclear Reactors	56
Reactor Types	57
Fuel Assemblies	60

Each assembly contains 0.5 metric tons of uranium and produces as much electricity as 60,000 metric tons of coal in a conventional power plant

Nuclear Reactors

All commercial nuclear reactors operating today are based on nuclear fission.

A nuclear reactor is designed and built to recover the heat energy produced by the controlled fission of uranium and/or plutonium contained in the fuel assemblies in its core and to use it to heat water and make steam, at high temperature and under high pressure. The steam drives a turbine linked to an alternator, which produces electricity. The steam is then cooled down and the condensed water recycled in the reactor.

Apart from its energy source, a nuclear reactor operates just like a thermal power plant. In the latter, heat comes from the combustion of fossil fuel, such as coal, oil or natural gas.

As nuclear energy is much more concentrated than chemical energy, only very small quantities of fissile materials are needed to fuel a nuclear reactor: the core of a PWR is made up roughly 200 fuel assemblies, each of which contains around half a ton of enriched uranium. The fuel in each assembly produces power for about 4 or 5 years before it is spent and removed from the core. During this time, it will have produced as much electricity as 60,000 tons of coal in a conventional power plant.

As opposed to conventional thermal plants burning fossil fuels, a nuclear reactor does not release polluting gases or carbon dioxide in the environment, and releases a very small amount of radioactivity in normal operation. On the other hand, it produces solid radioactive waste in its core, which must be safely managed to protect man and the environment from its harmful effects.

How a Thermal Power Plant works

How a Nuclear Power Plant works

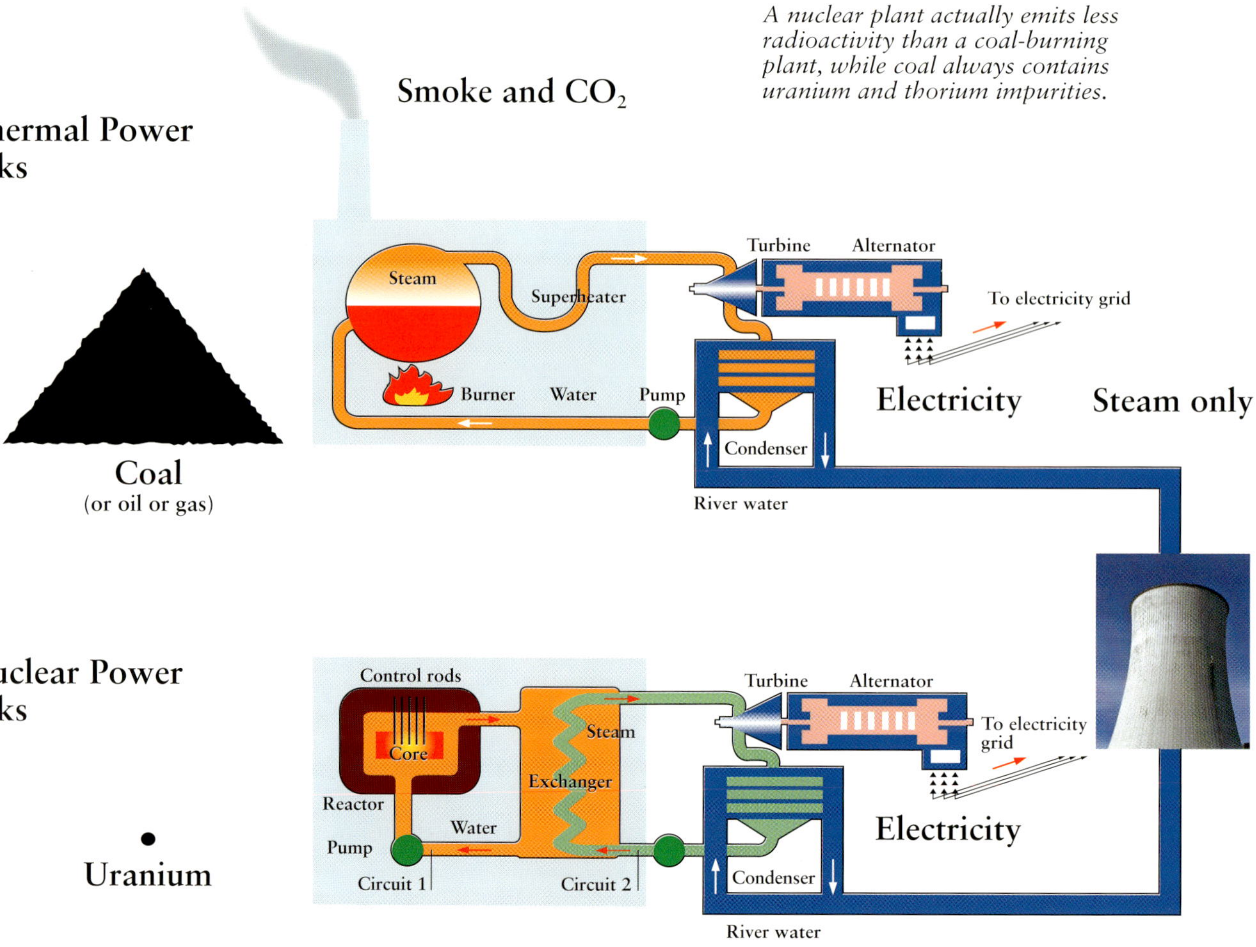

A nuclear plant actually emits less radioactivity than a coal-burning plant, while coal always contains uranium and thorium impurities.

Diagram of a PWR Plant (Pressurized Water Reactor)

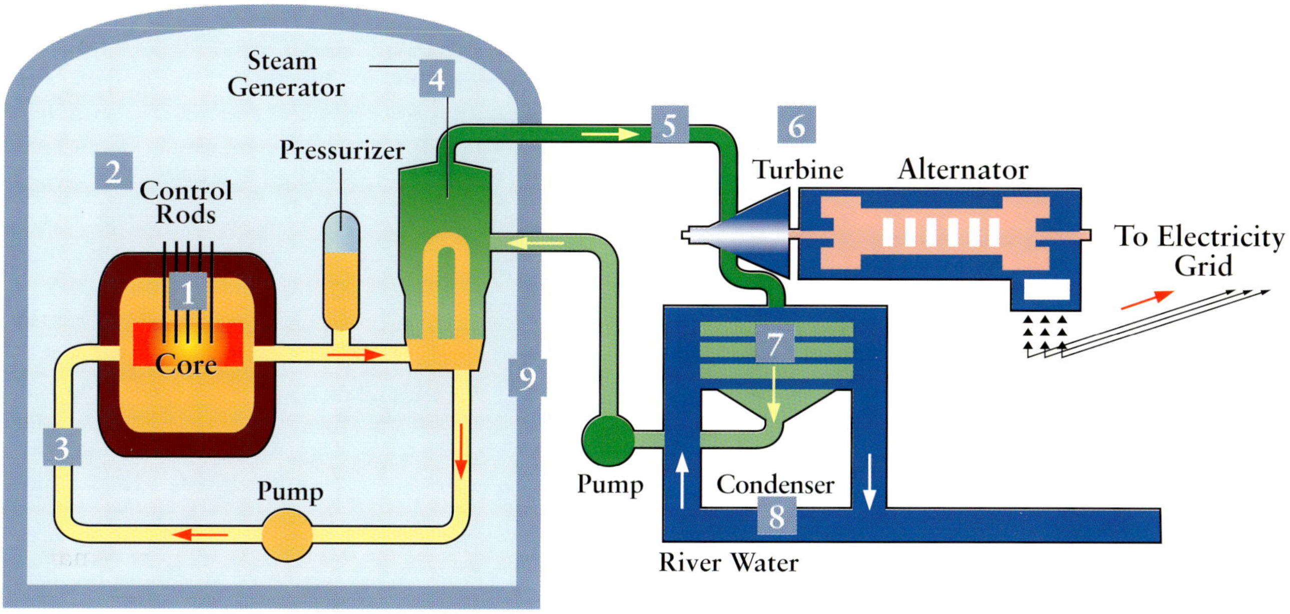

1. Fissions are produced in the core, which comprises a series of fuel assemblies. Each assembly is made up of long, sealed metal tubes, inside which are packed ceramic pellets of enriched uranium oxide.
2. The power released by the fissions is controlled by moving neutron-absorbing control rods in and out of the core, up to different depths.
3. The core is cooled by water, maintained in liquid form at 300°C by a pressurizer, and contained in a closed circuit within the thick-steel sealed primary circuit.
4. The primary water transfers its calories in a steam generator to the water of the secondary circuit, itself sealed. The secondary water is converted into steam.
5. The steam drives a turbine
6. The turbine is connected to an alternator.
7. The steam is allowed to condense in a condenser.
8. The condenser water is cooled down by a third circuit comprising cold water taken directly from a river or the sea or is sent to a cooling tower before being discharged into a river.
9. The primary circuit is encased within a containment building made of very thick concrete.

The diagram above shows a Pressurized Water Reactor, or PWR. PWRs and their close relations, the Boiling Water Reactors (BWR), make up most of the 440 nuclear reactors operating worldwide (see chapter.7). 58 PWRs were operating in France in 2006.

Reactor Types

PWRs and BWRs are both thermal neutron reactors, so called as the neutrons emitted at high speed (20,000 km/sec) by one fission are slowed down to about 2 km/sec before they cause the next fission. The neutrons are moderated through collisions with light nuclei, making them more efficient for maintaining the chain reaction.

In PWR and BWR reactors, the water for cooling the core is also used as the moderator and if the water is lost, the fission reactions will shut down by themselves. As the water absorbs some of the neutrons, the reactors need enriched uranium (which will be explained in ch.6).

There are many other types of reactor that were once used or are still operating now, and these include:

- Pressurised, carbon dioxide-cooled, graphite-moderated reactors using natural (i.e. unenriched) uranium as fuel. The design, known as UNGG in France and Magnox in Britain, is now obsolete

58 PWRs were operating in France in 2006, generating about 80 percent of the country's electricity...

Reactors moderated with graphite and cooled with ordinary boiling water. "RBMK"Chernobyl.

- Candu pressurized heavy water reactors, cooled and moderated with heavy water* and fuelled by natural uranium, designed in Canada.
- RBMK graphite-moderated reactors cooled by boiling water. The Chernobyl explosion in April 1986, the most serious nuclear accident to date, involved an RBMK reactor.
- High Temperature Reactors, HTR, are advanced versions of graphite-moderated nuclear reactors and produce very high core-outlet temperatures that offer the prospect of cogeneration, producing, for example, hydrogen and electricity at the same time.

In Fast Breeder Reactors, FBR, the neutrons are not slowed down between fissions. They need a larger quantity of fissile material to operate, but they use this material far better**. Considerable research is being carried out on breeders to design the next generation (IV) of nuclear reactors (see chapter 12).

** The nucleus of ordinary hydrogen is made up of a single proton. Deuterium is another form of hydrogen atom, but with a proton and a neutron in its nucleus. Water containing deuterium atoms is called "heavy water" and is a better moderator than ordinary water, but expensive to produce.*

*** It actually makes more efficient use of the fertile material. A kilogram of natural uranium can provide nearly 100 times more energy in an FBR than in a PWR.*

Comparison of Reactor Types

Type	GWe 2005	Fuel	Moderator	Coolant	Core Volume Power (MWt/m^3)	Thermal Efficiency (%)	Comments
UNGG, Magnox	2.3	Natural Uranium metal rod	Graphite	Pressurized CO_2	1	31	Only a few Units left in the UK
AGR	8.4	2% enriched Uranium oxide rod	Graphite	Pressurized CO_2	2	41	Only in the UK
PWR, VVR	240.6	4% enriched Uranium oxide rod	Ordinary Water	Ordinary Water	100	34	Derived from Submarine Technology
BWR	82.6	4% enriched Uranium oxide rod	Ordinary Water	Ordinary Water	50	34	
Candu	20	Natural Uranium oxide rod	Heavy Water	Heavy Water	10	29	No Uranium enrichment
RBMK	12.4	2% enriched Uranium oxide Rod	Graphite	Ordinary Water	5	33	No Future after Chernobyl
FBR	1	Uranium/Plutonium oxide Rod	None	Molten Sodium	500	42	Allows breeding
HTR	0	Coated Particles in graphitized Matrix	Graphite	Pressurized Helium	10	> 45	Allows co-generation

Assembly of PWR fuel pins.

Arrival of a PWR steam generator at EDF's Saint-Laurent Plant.

Fuel being loaded in the core of a PWR.

Fuel Assemblies

Nuclear fuel is expected to fulfil the following functions:

- Providing the power required by the reactor for the entire duration of the planned cycles, while adapting to the power variations imposed by the grid.
- Containing fission products under normal, incident and accident operating conditions (within design limits).
- Achieving the lowest possible cycle cost by optimizing the use of available fissile material.

The above objectives generate technical and technological requirements, of which the most important are:

- High power-density: a PWR fuel rod (9.5 cm diameter) produces on average 180 W per cm length, which means that 10 cm of fuel rod produces as much heat as an electric iron. The power produced in 1 liter (1 dm^3) of PWR core is 100 kW, i.e. the same power density as a car engine.
- High level of reliability combined with a long lifetime: the fuel assembly, its structure and the rods comprising it need to operate without failure throughout their time in the reactor, currently 4 to 5 years, with a target of 6 years by 2010. The current annual failure rate is equivalent to a slight cladding leak in one rod out of a million. That translates into one rod failure per year, on average, for all French reactors put together. The objective is zero defect.
- Leaktightness: in incident and accident situations, nuclear materials must for safety reasons remain perfectly contained. The fuel rod cladding forms the first barrier to contain the radioactive fission products (the other two barriers are the primary system and the containment building).

Under extreme accident conditions, even if cladding failures are unavoidable and the fuel assembly is deformed, it must still be able to cool the fuel and evacuate the residual heat (the heat which is still released after the reactor is shut down.

In spite of these performance requirements, the fuel design must remain simple, and this need for simplicity applies to all stages, i.e. manufacturing, handling, transportation, repair (defective rods can be replaced during operation), and, after use, storage.

Another general requirement is that it must be possible to reprocess spent fuel to recover recyclable nuclear materials, such as uranium and plutonium, and also to efficiently condition radioactive waste. These measures help reduce costs, provide radiation protection and contribute to the preservation of resources.

Another general requirement is that it must be possible to reprocess spent fuel to recover recyclable nuclear materials, such as uranium and plutonium, and also to efficiently condition radioactive waste

Finally the choice of materials for the structure of the assembly and cladding must take into consideration their resistance to irradiation and corrosion, to reduce as far as possible the amount of activation products (radioactive material resulting from the bombardment of stable substances by neutrons, typically around the core) which contaminate the reactor primary circuit, the fuel assemblies themselves, and thus the pools where the spent fuel is stored.

Description of a PWR fuel Assembly

The fuel assembly of a light water reactor is always made up of rods containing the nuclear materials. These rods are arranged in a square grid in a structure designed principally to provide mechanical support for the fuel-rods. In a pressurized water reactor, this structure is open and the coolant can flow across the assemblies.

All assemblies intended for PWRs are made up of a structure comprising a lower and upper end fitting, and guide tubes to which the fuel-rod spacer, grids are attached. The absorber (or control) rods of the rod-cluster control-assembly slide into the guide tubes. One guide tube is reserved for core instrumentation.

A Fuel Assembly

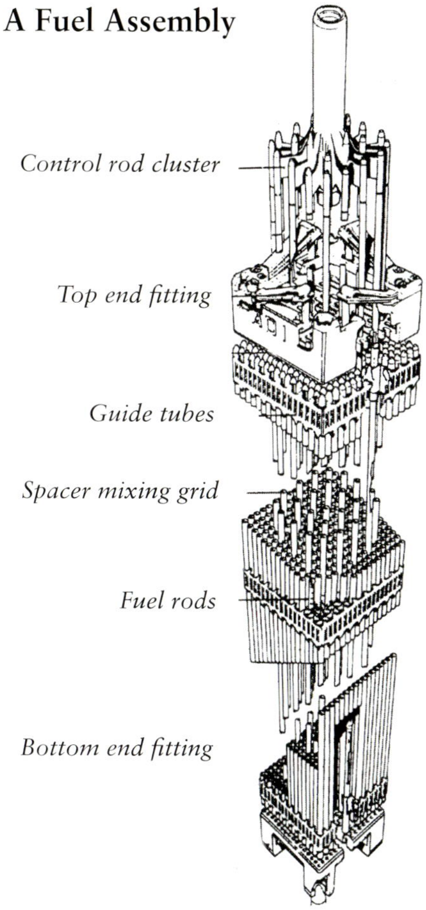

Fuel rods are made up of a leaktight cylindrical cladding of zirconium alloy in which are piled up ceramic pellets of enriched uranium oxide (or of a mixture of uranium oxide and plutonium oxide).
The rods are initially filled with pressurized helium to balance the pressure of the primary circuit water. An empty plenum is kept on top of the rod, in which the gaseous fission product can accumulate.

Nuclear Reactor timeline from 1942 to now

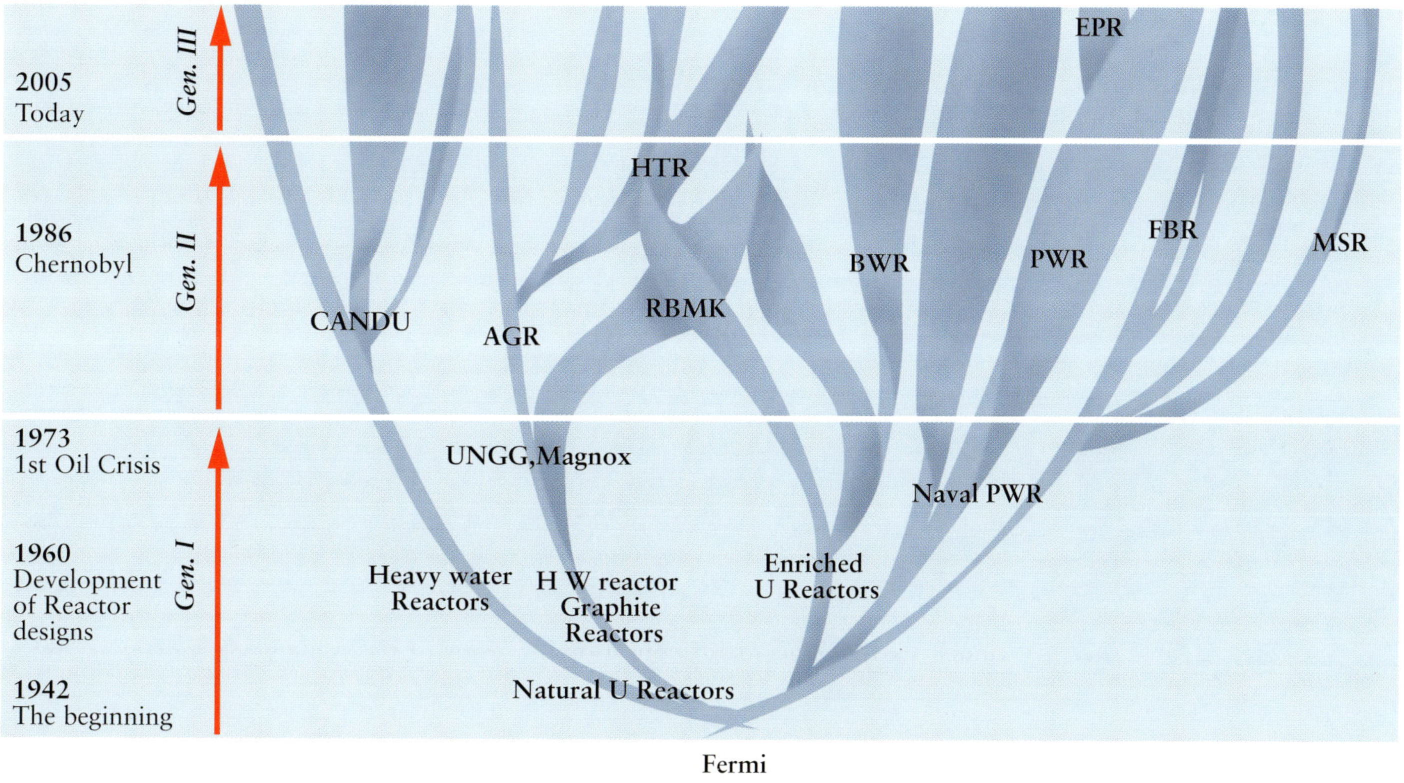

The PWR Fuel Cycle

The Fuel Cycle and its Different Stages	64
Once-Through Cycle and Closed Cycle	65
Uranium Extraction	65
Uranium Resources and Reserves	69
Cigar Lake, a Monster Deposit	70
Isotopic Enrichment of Uranium	73
Georges Besse Plant: Eurodif	74
Fuel Manufacture	76
Spent Fuel Storage	77
Spent Fuel Reprocessing – Recycling	78

The Fuel Cycle and its different Stages

According to the standard definition, a fuel is a substance that can be combined with oxygen to produce heat by combustion. By extension, the term nuclear fuel is applied to an element that can produce heat through the fission of the heavy atoms it contains.

We do not take oil as it gushes from the well and pour it directly into the gas tank of our car, nor do we connect electric light bulbs directly to a hydroelectric dam. In the same way, uranium ore does not provide nuclear fuel directly. For heavy nuclei to generate usable heat through fission, they have to undergo a fuel cycle consisting of a number of industrial stages, outlined in the following diagram:

The Fuel Cycle

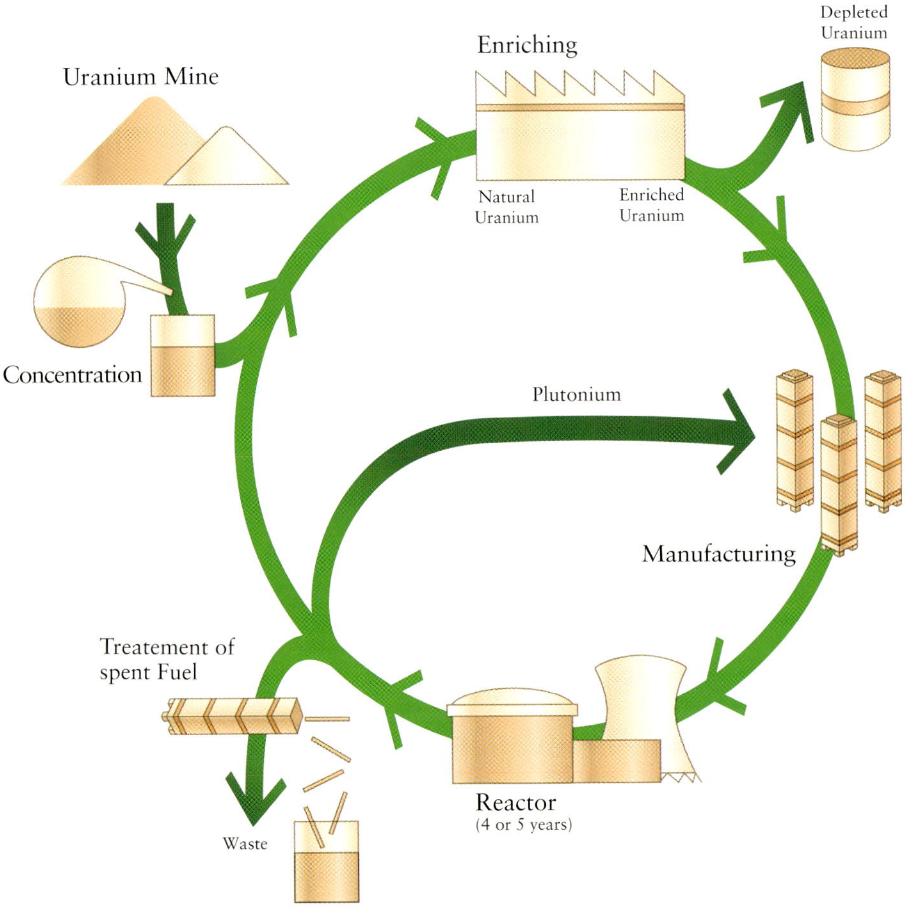

Fuel cycles depend on the type of reactor and the choice of fissile and fertile isotope pairings.

The typical PWR fuel cycle is made up of the following stages:

- Extracting uranium ore from underground mines, open quarries or in-situ leaching,
- Concentrating the ore (which often contains less than 1% uranium) on the extraction site,
- Converting uranium concentrates into uranium hexafluoride, UF_6, which is solid at ambient temperature and vaporizes at moderately high temperatures,
- Isotope enrichment of UF_6 to increase the concentration of fissile ^{235}U nuclei, as it is too low in natural uranium,
- Fuel manufacture (i.e. conversion of the fluoride into enriched uranium oxide, UO_2, pellet manufacture, pellet sintering, fuel rod manufacture, and assembling the rods into bundles),

Loading Fuel in the Reactor Core

↓

Electricity Production in the Reactor for about 4 years

↓

Spent Fuel unloaded from the Core

- Interim storage of the spent fuel under water,
- Spent fuel management and possible recycling of recoverable materials.

There are many inspections and some transportation between each of the above stages. Each stage is a complete industrial process in itself, as shown above with the manufacturing stage.

Each installation involved in the fuel cycle, i.e. enrichment, manufacturing or reprocessing plant, is large enough to produce fuel for several dozen large reactors. The fuel cycle is international by its very nature.

Once-Through Cycle and closed Cycle

The last stage - spent fuel management - is different for the closed cycle, set out in the diagram above and adopted in France, and the once-through cycle, adopted notably in Finland and Sweden.

The closed cycle includes the following sub-stages:

- Chemically processing the spent fuel to recover any remaining fissile and fertile materials for recycling purposes,
- Recycling the plutonium into MOX fuels and enriching residual uranium,
- Conditioning waste, in particular, vitrifying high-level radioactive waste left over from fission,
- Final disposal of conditioned waste.

The once-through cycle - which in fact is not a cycle at all - ends with the final disposal of the spent fuel as a whole, in this case written off as ultimate waste.

Uranium Extraction

Isotopic Aspects

Uranium posseses many isotopes, all of which are radioactive. All these isotopes were formed at the same time as the Solar System, but only those with a long-enough half-life are still to be found in the Earth's crust. The only ones still present in significant quantities are the isotopes of atomic mass 235 and 238, symbolized ^{235}U and ^{238}U.

Isotope	Half-life (years)	Current relative abundance on earth (as % total U)
235	713 million	0.72
238	4.47 billion	99.275

As can be seen in the table, natural uranium consists of 99.3% ^{238}U and 0.7% ^{235}U, and there has been no physical or chemical process in the natural environment that has led to any significant split of the two isotopes.

The radioactive half-life of ^{238}U is much longer than that of ^{235}U, so the $^{235}U/^{238}U$ ratio has been going down throughout the Earth's history. The result is that 4.5 billion years after our planet was formed, half the ^{238}U remains but only 2% of the ^{235}U.

Uranium Resources and Reserves

The insufficient availability of fissile material reserves is often quoted as one of the factors which could someday limit any large-scale development of fission nuclear energy.

This idea, spread in many papers, is fundamentally wrong. The authors usually simply make do with analysing only currently-operating reactors and the uranium issue and tend to lump uranium and hydrocarbon reserves in the same bucket. The issue of metal reserves, such as iron, copper, zinc, lead, gold and uranium is, however, very different to that of hydrocarbons. The explanation comes down to geology. Metals tend to be found underground in continuous veins and the reserves therefore fluctuate considerably with demand and prices, while oil and gas is found in discontinuous fields (with some exceptions, such as oil shale). This means that hydrocarbon reserves are not very sensitive to oil and gas prices, as was seen when prices rose sharply in the 1973 and 1979 oil crises.

Poor understanding of these questions has led many economists to excess optimism concerning the issue of oil reserves and to unfounded fears about uranium reserves. Geologists know that reality runs contrary to those visions.

Apart from the geological factors relating to uranium deposits, which mean reserves increase when prices go up, there are several other issues that show why raw materials for the nuclear industry need to be considered separately from oil and natural gas reserves.

The three main reasons are as follows:

- The energy efficiency of a given type of reactor, such as PWRs and BWRs, can

Geologically, uranium, like most metals, can be found in concentrated (ore pockets, veins) or diffuse deposits, usually in sedimentary strata: clay or shale, phosphate, limestone and especially sandstone.

The exception is the Oklo deposit, where natural nuclear reactions took place almost 2 billion years ago, consuming uranium-235 and upsetting the isotopic composition of the uranium remaining today.

Martin Klaproth

1789: The Silent Uranium Revolution.

Uranium was first described by Martin Klaproth, who presented the new substance in a paper to the Science Academy of Prussia, in Berlin, in 1789.
He thought he had discovered a new element and called it uranium in the honor of the planet Uranus discovered a few years earlier in 1781.
The substance was, in fact, uranium oxide extracted from pitchblende veins in the Joachimsthal region, which would become Jachymov in 1918 when Czechoslovakia was created. The region lies in the middle of the "Erzegebirge", the "Mineral Mountain", an old mining area in Bohemia-Saxony which had been producing silver, along with tin, lead and zinc since the Middle Ages.
Uranium was first isolated in its metallic state in 1841, by Frenchman Eugène Peligot.

Eugène Peligot

be improved by using richer or more sophisticated fuel, or by reprocessing spent fuel (see pages 78-80).

- There are other reactor types which, with the same amount of natural uranium, will be able to extract up to 100 times more energy than can be produced in less efficient fission reactors (see breeders chapter 12).
- Thirdly, uranium is not the only raw material that can be used, there is also thorium, which is probably more abundant than uranium, but does not have a fissile isotope.

The Issue of Uranium Reserves

Uranium is a commonly-occurring metal, found both within volcanic and crystalline rocks and in sedimentary-metamorphic deposits. It was first discovered in 1789 as an oxide and was extracted in metal form in 1841 (see margin note). It occurs in fairly low concentrations of about 3 ppm (or 3g / tonne), about the same as tin and molybdenum. This average abundance does not actually mean much as it varies considerably according to the host rocks. Granites and shale, for example, often exhibit abundances of around 15 to 20 ppm.

An important characteristic of uranium is that, contrary to thorium, it is soluble in water. Seawater therefore contains measurable quantities, amounting to about 3 ppb (parts per billion), in other words about 1,000 times less than the average abundance in Earth's crust (see page 69).

Ores and reserves, whether of uranium or any other mineral resource, should be considered in terms of economics. They are deposits that can be exploited economically under current price conditions (of uranium or other metals) and using existing production technologies. Uranium ores (2000/2005) currently being mined contain a wide range of uranium concentrations.

As uranium is water-soluble, it will be dissolved through the erosion and leaching of uranium-bearing crystalline or metamorphic rocks, later to be trapped in sedimentary rocks near the eroded crystalline zones. The geological media where the dissolved uranium is most likely to be trapped include sandstone rich in organic matter or other sedimentary reducing zones. These sedimentary traps can contain highly variable amounts of uranium.

Uranium Prospecting

The first task in prospecting a sector, whatever its size, is to gather all the available geological and mining data available. Today, the use of satellite imaging and geophysical prospecting techniques has considerably simplified the detection of potentially favorable zones.

Radiometric prospecting is the geophysical method used specifically for uranium detection, as it is based on radioactivity. The technique actually ends up detecting more radium, a daughter product of uranium, which is over a million times more radioactive than uranium. Radiometric prospecting can be performed on foot or from a land vehicle or aircraft for large grids and regions where access is difficult.

Although correctly-interpreted radiometric measurements can be used to detect uranium ore deposits, they give little information as to the quantity or concentration of the ore. Additional information can be obtained through in-situ electrical, electromagnetic and magnetic measurements. If the results are encouraging, a more in-depth survey may be made by drilling holes to allow direct access to the ore.

The table on page 70 illustrates another major difference with the coal and gas industries, where every year consumption and production are equal to within 1% (due to stockpiling and stockdrawing).

The relationship between "military nuclear" and "civilian nuclear" (see chapter 4) may appear surprising. While many people are afraid that countries acquiring civilian nuclear reactors may someday

Underground or strip mines? In France, strict rules set out under Napoleon in 1810 regulate the two methods.

"Jumbo" hammer drill used deep in the Jouac mine (France), which was closed down in 2002.

World Uranium Production and Consumption

	2000	2005	2020*
Consumption in Reactors**	60,000 t	67,000 t	(90,000 t)
Production from mining *	35,000 t	45,000 t	(75,000 t)
Secondary Sources**	25,000 t	22,000 t	(15,000 t)

** Authors' estimation*

*** Reprocessing, stockdrawing military stockpiles during this period of decreasing stockpiles, uranium prices remained so low that mining companies were not in a position to invest normally in exploration or exploitation of identified resources. The prospective decrease of those secondary sources together with the now anticipated acceleration of nuclear programs have provoked spectacular price increases since 2004 (see graph page 71). These price increases (x 3 in two years) have triggered renewed new exploration efforts.*

use their facilities (especially enrichment facilities) or their nuclear materials (enriched uranium and plutonium) to acquire nuclear weapons, what we see here is the contrary. Civilian nuclear power contributes to disarmament by consuming uranium or plutonium stockpiles which had been accumulated for military purposes.

Let us not forget either that current reserves are estimated on the basis of $130/kg. If prices rose to $200/kg, reserves would approximately double within a few years.

Phosphate ores contain more uranium than conventional deposits, and extraction processes have been developed on an industrial scale.

There is also thorium, which will be considered later.

Reserves expressed in tons or billion kWh?

Even more important than estimating the uranium reserves is assessing the amount of energy they can supply. Here there is no comparison with hydrocarbon and coal reserves. If you improve thermal reactor technology, you might gain an extra few tenths of one per cent yield. However, a switch from current nuclear reactors to breeders can produce about a hundred times more energy from the same quantity of uranium, not to mention the gains in thermal efficiency.

Breeders can also be used to extract the 99% of potential energy left in the by-products of PWR fuel, such as the depleted uranium stored after the enrichment of natural uranium and the uranium recovered from spent fuel reprocessing.

2,000,000 tons of depleted uranium are already stockpiled in the world. In breeder reactors they could produce 20 million TWh, or more than 1000 times world electricity consumption in 2005!

Nuclear Power and Security of supply

It is sometimes argued that as uranium is imported, there are the same problems of security of supply as for oil and gas.

This is not true, as stockpiling fuel for producing electricity is much cheaper for uranium than for coal, gas and oil: uranium storage works out some 30 times cheaper than for oil (2006 conditions)!

Main Uranium Deposits

Deposits are areas in the Earth's crust where local physicochemical conditions have caused uranium to accumulate and reach concentrations of over 0.1%, although this can also reach several dozen % in certain areas (see box).

In most cases these deposits were formed more than 600 million years ago during the Precambrian era.

World Resources and Reserves

The table below gives an estimation of world uranium resources in 2006, in million tons. Can these reserves be described as extensive? For the sake of comparison, 2 million tons of uranium have been produced since the dawn of the nuclear industry, almost the same amount as the current estimation of RARs.

Uranium from Seawater: Bonanza or Myth?

Seawater contains on average about 1000 times less uranium than terrestrial crust, but about 10 times more in the form of "very low-grade potential reserves". The terrestrial crust contains around 3 ppm of uranium (3 g/t) and seawater slightly more than 3 ppb (3 mg/t), but if you multiply this concentration by the total mass of all the seas and oceans, you can see that they contain around 4.5 billion tons of uranium, about a thousand times more than currently-identified terrestrial reserves (4.5 Mt).

Seawater might therefore in the long run supply uranium resources, because while it is a very low grade "ore", uranium is already present in ionic form. Tests in Japan have already shown it is chemically feasible to recover uranium from seawater, using ion exchange resins. The main problem is the volume of water that needs to be filtered to recover industrial amounts of uranium, while avoiding pumping water that has already been filtered. One possible solution could be to build huge floating facilities anchored in the middle of the main ocean currents, such as the Gulf Stream and the Kuro Shio. The size and cost of these giant structures, which need to withstand everything the ocean can throw at them, has yet to be worked out.

Seas and oceans contain about 4.5 billion tons of uranium, about 1,000 times more than currently-identified terrestrial reserves (4.5 Mt)

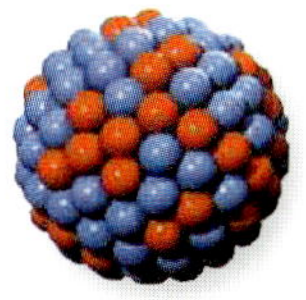

At the current rate of consumption (the world need for natural uranium in 2005 was about 67 000 tons), reserves (RAR and inferred) recoverable at less than 80 $/kg U, i.e. 3.8 million tons, should last more than 55 years.

The "ultimate" reserves, perhaps 19 million tons, would last well over two centuries. Beyond this horizon, the millions of tons of uranium contained in phosphate ores could still be extracted. On the other hand, if consumption grows rapidly, the duration of reserves will be shortened accordingly with the same reactor technology.

Million metric tons Recovery Cost Ranges	Reasonably assured Resources (RAR)	Inferred Resources	Forecast Resources	Speculative Resources
< $40/kg U	1.95	0.80		
< $80/kg U	2.64	1.16	1.7	
<$130/kg U	3.30	1.45	2.5	4.6
Unknown				7.5

(Lines cannot be totaled, with the exception of the last one)

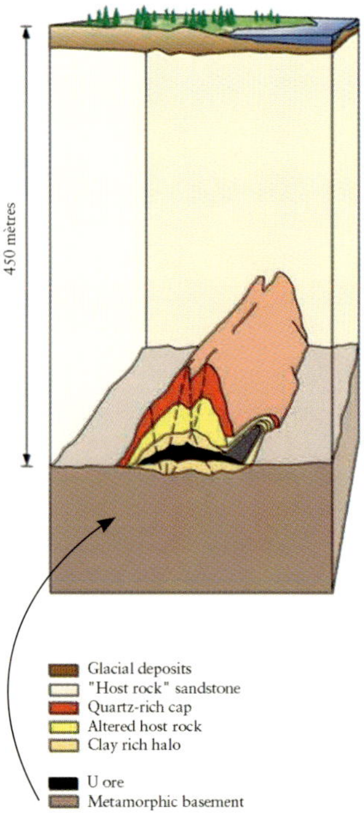

Cigar Lake, a "Monster" Deposit !

The Cigar Lake deposit lies in the Canadian Athabasca region. There were no surface indications of the existence of this uranium oxide lode, the largest ever discovered. It was only discovered through the search, mostly geophysical, for structures analogous to deposits previously discovered in the region.

The deposit, the content of which is put at 150,000 metric tons of uranium, is made up mostly of uraninite, with grades ranging from 8 to 55%. It was formed at a depth of 3,000 meters, 1.3 billion years ago, at the interface between the sandstone and metamorphic rocks of the Canadian Shield. Hydrothermal reducing fluids percolating through the base rock provoked the precipitation of uranium in the sedimentary basin above.

The same hydrothermal fluids, by dissolving silica from the sandstone above the deposit, caused the formation of a clay lens which helped protect the ore after its formation.

After thousands of years of erosion, the deposit is now only 450 meters below the surface. Its size is close to that of the Champs Elysées in Paris!

Reasonably Assured Resources: Main Countries - 2005

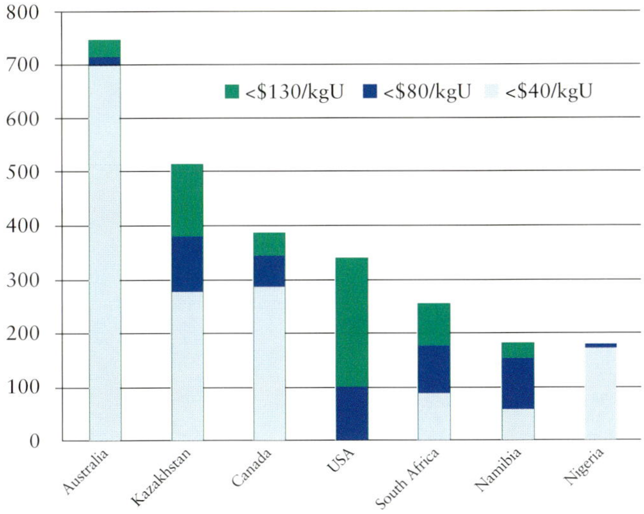

Inferred Resources

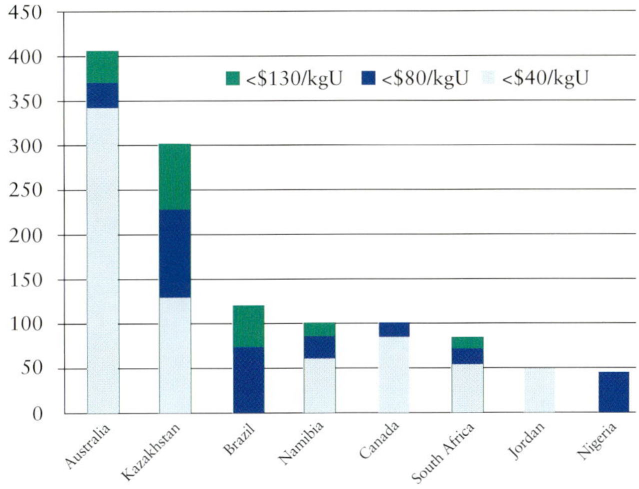

While needs amounted to 67,000 t in 2005, production reached only 45,000 t, around 60% of the consumption. The remaining part was supplied from secondary sources: stockpiles, re-enriching slightly depleted uranium, uranium recovered from spent fuel reprocessing, etc... Those secondary sources will decrease quickly, but their abundance in the past decade had badly depressed uranium prices* and many mines had to close down. Since 2003, uranium prices have experienced a steep increase, and prospecting is back with a vengeance.

What do uranium mines look like?

In order to extract the uranium, access must be gained to the deposit, either by stripping off the sterile rock covering it (as in the case of open-pit mining), or by digging galleries into the sterile rock if the ore lies deeper (as in the case of underground mining), or by injecting into the rock a solution which dissolves the ore in situ and which is recovered at the surface. This process is called "in situ leaching".

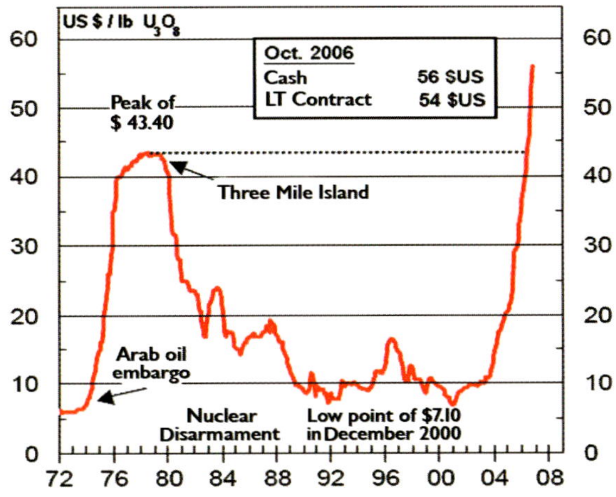

The advantage of open-pit mining is that the ore is easily accessible. The techniques and equipment used are similar to those employed in quarries or public works sites. The rock cover is stripped off by a bulldozer, and then the ore is extracted by a mechanical digger, and hauled by truck up a ramp to ground level. Once all the ore has been extracted, the excavation can in some cases be flooded to form a lake. In most cases, however, the hole is filled in with the excavated material. The topsoil layer is then put back into place to create a suitable landscape.

In-situ leaching,

This is a technique used to extract deep-veined uranium without having to dig extensive underground structures. A number of wells are drilled to access the ore and then a solution, usually alkaline, is injected into the wells to dissolve the uranium, which can then be recovered by pumping it up through neighboring extraction wells.

2004 Production
(40,000 tU)

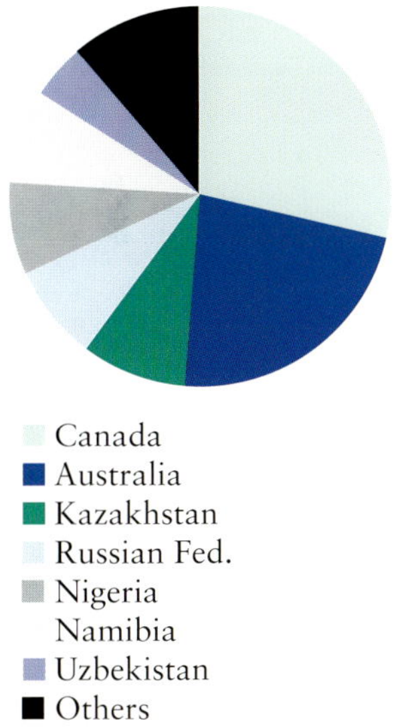

* *From 1989 to 2003, spot prices stayed close to \$10/lb U_3O_8. In 2006, they exceeded \$50/lb.*

An open-pit mine in Australia (Ranger mine).

Miner in a uranium mine in the then Czechoslovakia, 1972.

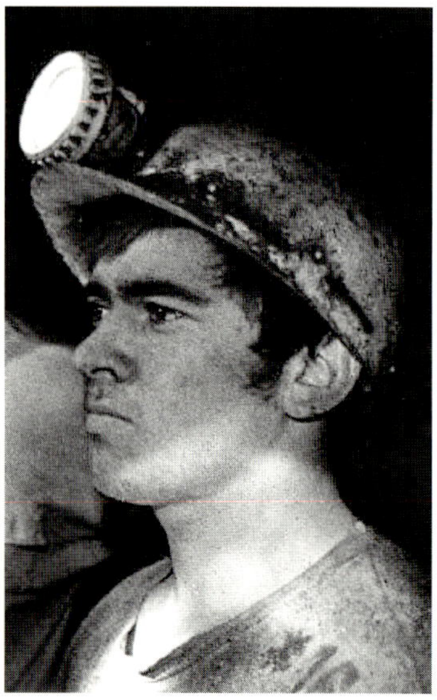

In-situ leaching is an economical method that can be used to exploit low-grade resources, even at considerable depths. Great care, however, has to be taken to avoid polluting underground water tables, and therefore is used only in sparsely inhabited regions (Wyoming, Uzbekistan, Kazakhstan, Australia).

At the moment, the ore deposits considered economically worth extracting contain at least one kilo of uranium (several kilos in some cases) per ton. Canadian ore deposits in Saskatchewan are high grade, with deposits containing up to and even over 10% uranium. French ore deposits are lower grade. COGEMA, the French nuclear reprocessing company, has extracted some 52 million tons of ore to obtain 72,000 tons of uranium.

Ore processing

As uranium deposits are often of low grade, the ore must be concentrated on the spot before transportation. This operation is performed in a concentration mill. After crushing and grinding, the ore is then chemically dissolved to form a pulp, in most cases using sulfuric acid. The yield is generally higher than 90%. After clarification and purification using solvents or ion-exchange resins, the uranium from the solution is precipitated by adding ammonia, magnesia, magnesium hydroxide or soda.

After washing and filtering, the concentrate contains about 75% metal uranium and is known as "yellowcake". This can be roasted to obtain uranium oxide U_3O_8. Yellowcake and uranium oxide are the marketable end-products of the processing plant.

Uranium Conversion: from Yellowcake to Hexafluoride

At the yellowcake stage, the uranium is still nowhere near ready for use in a nuclear reactor. It must first be converted into uranium hexafluoride or UF_6 (gaseous) to undergo isotope enrichment.

The first step is to purify the yellow cake to rid it of neutron-absorbing elements such as boron or cadmium, or other elements that form volatile fluorides liable to contaminate the uranium hexafluoride produced at a later stage. Unwanted elements are extracted using tributyl-phosphate TBP, after dissolving the yellowcake or uranium oxide in nitric acid.

The purified uranyl nitrate solution is then converted into uranium oxide powder, and this powder is then fluorinated in two stages and converted into uranium hexafluoride UF_6. The uranium hexafluoride is subsequently transferred to the enrichment plant in solid form in pressurized containers.

Yellowcake

Uranium Isotopic Enrichment

Enriching uranium means increasing the proportion of ^{235}U isotope, from the natural value of 0.7% to the isotopic concentration required to fuel light water reactors, which stands at around 4% today. In these reactors, the enrichment level is laid down by the use of light water as a moderator and coolant, and the need to compensate for neutron capture by the hydrogen nuclei in the water (see chapter 5).

The physical properties of a pure element depend on its isotopic composition, but these differences in properties are generally slight and not easy to exploit for the purposes of isotope separation. The processes implemented use the difference in the mass of the isotopes to be separated, either directly (mass spectrometry and centrifugation), or indirectly (gas diffusion). The chemical properties of elements generally depend very little on the isotope in question, which makes isotope separation all the more difficult by chemical means.

Cascades

Both industrial separation methods, gaseous diffusion and ultracentrifuge, have a small or very small elementary separating power, which calls for the use of a cascade design, where each elementary step is repeated until the required degree of enrichment, Np, is obtained (see diagram). The cascade is often made up of two parts, the larger one being devoted to the actual enrichment. Other stages are also used to deplete the residues in order to optimize the separation performance of the cascade.

The degree of depletion chosen depends on economic factors, including the price of natural uranium and the cost of enrichment. The higher the value of Nw, the easier the enrichment process, but the greater the consumption of raw materials. In order to achieve successful economic optimization, the notion of the separative work unit (SWU) is used. This is a unit of measurement incorporating isotope abundance and the 3 flow rates. The actual formula to compute the SWU is complex, but only the orders of magnitude need to be known.

The diagram shows typical values of these parameters for a typical PWR fuel.

A normal 900 MWe PWR would consume about 130,000 SWU/year, at a rate of 5 SWU/kg U, enriched to 3.7% from natural uranium.

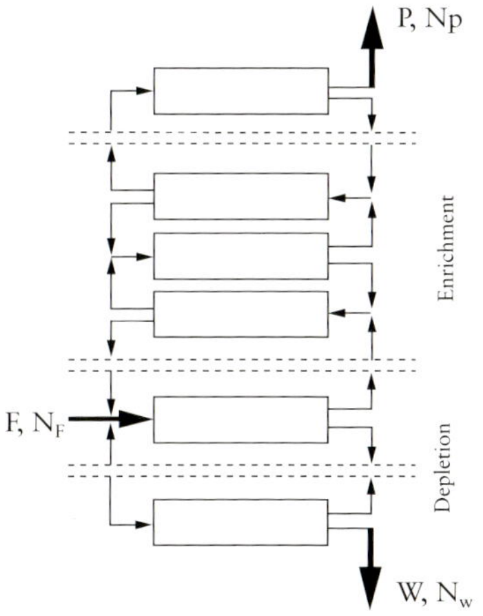

An example of a cascade. Each rectangle represents an elementary separation step. P(product), W(waste), F(feed) represent flows and Np, Nw Nf the isotopic abundances.

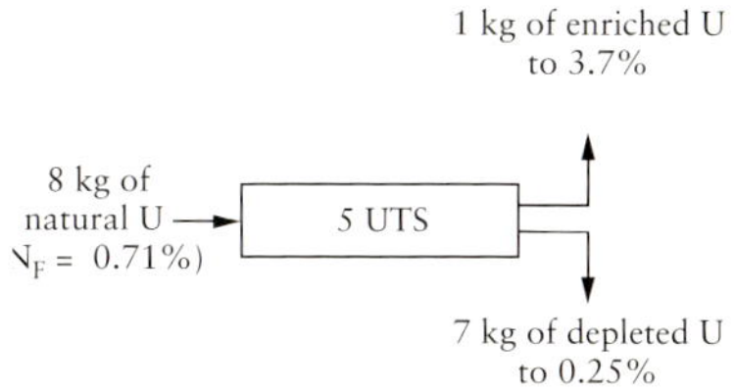

Orders of Magnitude: Enrichment of Uranium for Civil Use.

Gaseous Diffusion

Gaseous diffusion was the first uranium isotope enrichment process to be used on a large industrial scale. France has built two gaseous diffusion plants within its borders. The first was intended for purely

France's Georges Besse gaseous diffusion enrichment plant, in the backckground. In the foreground, the Tricastin nuclear reactors powering the plant.

military applications and has now been decommissioned. The second, for civil applications only, is the Georges Besse plant at Pierrelatte in the Rhône Valley.

The gaseous diffusion process for uranium enrichment uses a gaseous compound of uranium, uranium hexafluoride (UF_6). Fluorine has only a single isotope, of mass 19. The molecules of $^{235}UF_6$ (mass 349) travel faster than those of $^{238}UF_6$ (mass 352) and, over the same period of time, produce a greater number of impacts on the walls of the recipient containing them. If gaseous UF_6 is diffused through a porous barrier, then the ^{235}U enrichment of a fraction of the gas below the barrier will be slightly higher than for the total gas above the barrier at the start.

The elementary effect is small since the enrichment factor* for UF_6 is low, in practice around 1.002.

Important remark:
A gaseous diffusion plant has no flexibility whatsoever: once built, a cascade designed for low enrichment cannot be remodeled to produce military grade enriched uranium.
It is not, therefore very proliferation prone

Ultracentrifuge

One of the earliest designs the ultracentrifuge, is coming back into fashion. The process is statistical and operates in a gaseous phase: it involves rotating a cylinder containing UF_6 gas at very high speed. The heaviest molecules gather at the edge of the cylinder, while the lightest ones mostly migrate towards the center. The typical enrichment factor for this process is 1.3, a much higher value than that obtained with gaseous diffusion.

Depleted and enriched fractions are caught by scoops - small tubes fitted at both ends of the cylinder at carefully optimized distances from the wall (see figure opposite page). As the gas is fed in and caught by the scoops, a countercurrent forms along the rotational axis of the cylinder - which is vertical - and transforms the radial isotopic gradient into an axial gradient. This cross-current can be accelerated by heating the "bowl" to trigger convective vortices along the rotating wall.

With this optimized axial current, a centrifuge works rather like a distillation column: the current flowing upwards is gradually enriched with ^{235}U while the downward current is depleted.

Centrifuge operating temperature is governed by the properties of UF_6- gas temperature must not drop below the solidification point, as gas condensation on the

The Georges Besse Plant (EURODIF)

The Eurodif plant (Georges Besse plant), located at Pierrelatte in France, is intended purely for civil use. It was partially commissioned in 1978, with full-scale industrial operation as of 1982. The plant provides roughly a quarter of the SWUs produced each year in the West. The whole plant is basically made up of a single cascade with nearly 1,400 stages, of three sizes, mounted in series inside 4 large buildings.

Isotope separation by gaseous diffusion needs a great deal of energy, some 2,400 kWh per SWU. When running at full production capacity of 10.8 million SWU/year, the Georges Besse plant uses no less than three of the four 900 MW EdF reactors located on the same site at Tricastin.

The plant will be shut down around 2012.

** The enrichment coefficient is defined as the ratio of the enrichments N/(N-1) of the products before and after a unit stage.*

wall would create loss of mechanical balance which would in turn damage or destroy the machine. There are also a number of additional constraints: the material the rotating parts are made out of have to withstand both the enormous stresses induced by the centrifugal force and the corrosive power of the UF_6 gas. The main aspect in the design of high-performance centrifuges is coming up with a solution to overcome the problems of materials and mechanics.

State-of-the art centrifuges, with carbon fiber rotors, reach a unit separative capability ranging from 30 to 100 SWU/y. Even with the higher value, a 5 million SWU/y plant would consist of 50,000 machines running in parallel cascades, each cascade comprising 10 stages. The machines must be able to spin without stopping for several decades, as it is during the startup and slowdown phases that the risk of damage is the greatest. Because of the number of machines concerned and of the very long period of time without maintenance, the main challenge of a centrifuge plant is reliability.

With the today's high-performance materials, the centrifuge process is close to its technological peak and seems to have exhausted any possibilities of further progress in terms of performance and cost. The cost structure of a centrifuging plant is dominated by high investment costs and extremely low operating costs. Centrifuges are highly reliable in the case of models that have reached technological maturity. Centrifuge isotopic separation is close to a thermodynamically reversible process and as a result, the power consumption of centrifuging plants is very low, roughly 50 kWh per SWU, or 50 times less than a gaseous diffusion plant of the same capacity. Ultracentrifuge is presently the lowest-cost uranium enrichment process and it is set to replace gaseous diffusion over the coming fifteen years.

Other Processes

Historically, the first uranium isotopic enrichment process was electromagnetic separation, using mass spectrographs, which sorts individual ionized atoms according to their isotopic mass. This process has a low throughput and requires a lot of power (during the Manhattan Project, the Americans had to commandeer the silver reserves to make the coils for the magnets of the giant "calutrons" required to do the job). Everybody considered this process to be obsolete...until they found out in the First Gulf War that the Iraqis were using it!

In the early 90s, several countries (USA, France, Japan, South Africa) were putting a lot of effort into developing selective processes using lasers. The most notable of these processes was AVLIS, Atomic Vapor Laser Isotopic Separation, which basically involves the photo-ionization, or removal of an electron by laser light energy, of a ^{235}U atom by three photons, the added energy of which can ionize the atom. Selectivity between ^{235}U and ^{238}U atoms is possible as there are small differences between the forces binding the peripheral electrons, due to small variations in the volume of the nuclei. These differences in volume are due to the changes in the numbers of neutrons in the nuclei. Only lasers have wavelengths accurate enough to be used here.

Metallic uranium is therefore evaporated and laser beams tuned to the desired wavelength are sent through the uranium vapour. An electron is removed from the ^{235}U atoms, which then become positively-charged ions which are subsequently captured on negatively-charged metal plates placed in the vapour just above the laser beams.

All this is fine on paper, but much more difficult to put into practice. The French and the Americans who were working on the project managed to demonstrate its technical feasibility, but development was terminated as there was little chance of the process becoming economically viable in the short term. Perhaps its chance will come again in a few decades.

Also worth mentioning is the MLIS (Molecular Laser Isotope Separation) pro-

Simplified Cross-Section of a Centrifuge

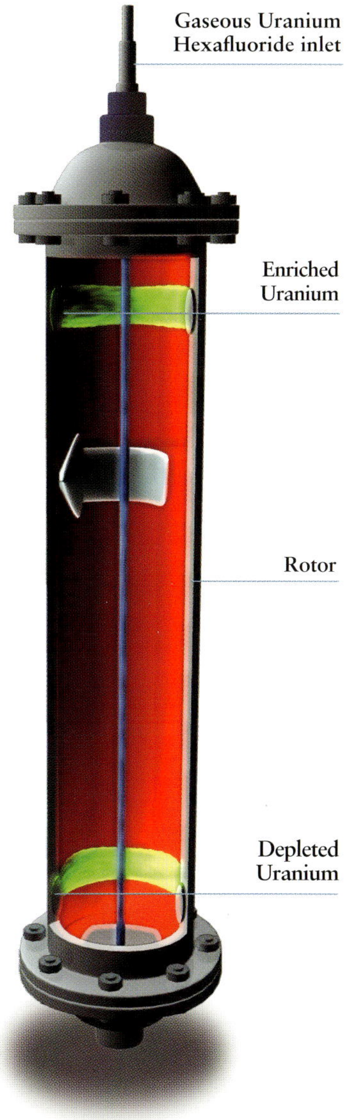

Cascades of Centrifuges. URENCO Company.

The ultracentrifuge enrichment technology may be deemed "proliferating" since it is not difficult to build small unobtrusive facilities. In addition, cascades designed initially for a low enrichment level can be rearranged to produce highly enriched, weapons gradt uranium. Here lies the crux of the present (2006) Iranian crisis.

cess, in which infrared lasers are used to selectively excite $^{235}UF_6$ molecules within a gas stream, where the UF_6 is mixed with another carrier gas, at low temperature. The excited molecules are photo -dissociated to UF_5 and fluorine. The process was abandoned in the mid-80s in the USA and France, before its technical feasibility had been fully demonstrated.

The Enrichment Market

There is presently (2006) an overcapacity of enriched uranium production in the world as, following the two oil crises, electricity demand did not follow the high growth scenarios of nuclear power development. This overcapacity situation has worsened as a result of the commercialization of military stockpiles.

Annual consumption is 32 million SWU/y. Production capacity is 45 MSWU/y, but actual production is only 30 MSWU/y. In addition, dilution of the 500 metric tons of highly enriched uranium that the USA bought from Russia, amounts to the equivalent of 5 MSWU/y, over 20 years.

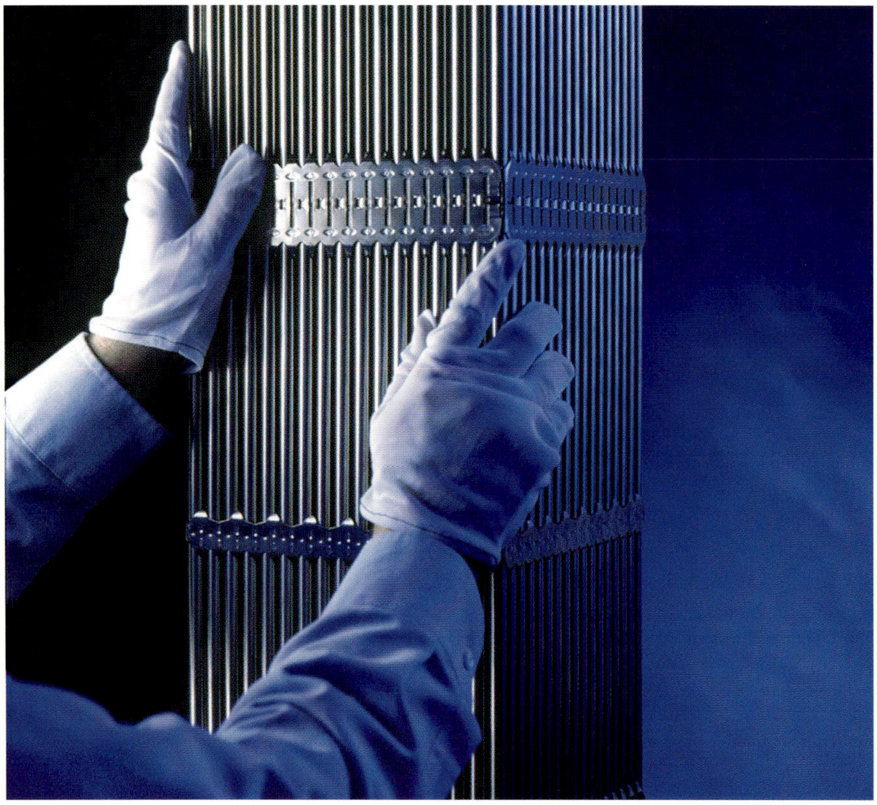

Fuel Manufacture (PWR)

You will remember from chapter 5 that the fuel assembly of a light water reactor is made up of rods containing the nuclear material. These rods are arranged in a square grid in a structure which provides mechanical support for the fuel rods, and which comprises a lower and upper end fitting, and guide tubes to which the fuel rod spacer grids are attached. The absorber (or control) rods of the rod cluster control assembly slide into the guide tubes.

Almost all types of fuel for light water reactors - PWR or BWR - are made from enriched uranium, containing 3% to 5% ^{235}U. France, Belgium, the United Kingdom, and soon Japan also make mixed oxide or MOX fuel. This is made using UO_2-PuO_2 and offers the advantage of allowing plutonium to be recycled.

The only differences between the manufacture of MOX fuel and standard UO_2 fuels lie in the preparation of the powder before pellet manufacture, and in the specific provisions made relating to the use of plutonium: radiological protection, containment, remote control, and inspections.

After defluorinating the uranium (UF_6) to convert it into oxide (UO_2), the actual fuel manufacturing process includes the following steps:

- Manufacturing the materials, chiefly zirconium alloys, and parts, tube cladding and structural parts.
- Preparing powders (UO_2) or UO_2 and PuO_2, including powders recovered from previous manufacturing runs.
- Pellet manufacture: shaping cylindrical pellets and sintering them, i.e. baking the ceramic.
- Rod manufacture: inserting the pellets and sealing the rods.
- Assembly and storage.

Each of these steps involves controls to ensure compliance with physical criteria (geometry, density and appearance), chemical criteria (exact degree of oxidation,

Pool storage, in The Hague, for spent fuel assemblies before reprocessing.

impurities, etc.), and isotope criteria (^{235}U abundance, and plutonium abundance). Quality assurance also requires materials, equipment and operations to be fully traceable so that the history of each item can be reconstructed and its origin guaranteed.

PWR and BWR Fuel Fabrication Market

The yearly global market for light water reactors represents around 6,000 tHM*: 4,000 tHM for PWRs and 2,000 tHM for BWRs. Total manufacturing capacity is far greater, and durably so. Firstly, yearly requirements are going down as burnup values go up. Secondly, UO_2 fuel manufacturing plants do not require heavy investment and their maintenance costs are low, so there is little incentive to reduce capacity.

MOX fuel markets are limited in volume: only the plutonium separated during reprocessing is recycled. Current demand is around 250 tHM/year.

The French market is about 1,100 tHM/y, of which 100 tHM/y is MOX fuel.

Spent Fuel Storage

Once unloaded from a reactor core, spent fuel assemblies are first stored on site in the pool of the Fuel Building, where they are kept for a minimum of 12 months. This storage time allows the spent fuel to cool down significantly. If they are bound for reprocessing, the fuel assemblies are then shipped, dry, in massive shipping casks to the storage pool, which is the head-end of the reprocessing plant (see above).

If they are not bound for reprocessing, they can be put inside huge dry-storage casks similar to shipping casks,

* tHM= ton of Heavy Metal: tonnage of uranium (and plutonium) contained in the fuel.

Shipping casts for 12 spent fuel assemblies.

or shipped to central storage facilities, maybe dry halls or pools, at the surface or underground at shallow depth (Sweden).

Spent Fuel reprocessing-recycle

What does spent fuel contain?

When spent fuel is removed from the core, it is still made up of about 95% uranium, much less enriched than it was initially and slightly "poisoned" by other non-fissile isotopes, about 1% plutonium, various fission products and a few minor actinides. The residual uranium and plutonium can be extracted and recycled to produce more energy. The remaining products are unusable ultimate waste.

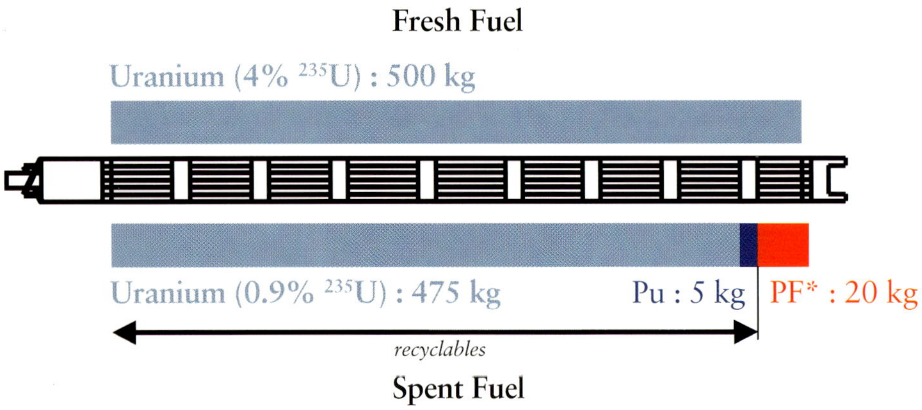

(and a few minor actinides)

Reprocessing the spent fuel means separating the various components so that each category can be treated individually. As the material handled is highly radioactive this makes it a heavy and costly operation. Uranium, plutonium (perhaps minor actinides too, in the future), and all highly radioactive fission products are separated in this way. Once isolated, they can be stored in suitable conditions, bearing in mind that most of their radioactivity will have disappeared in a few centuries. The fission products are bonded within a glass matrix, specially formulated to house such complex mixtures. The glass corrodes very slowly and can safely store the radioactive products for a very long period of time (many thousands of years).

Uranium presents no particular problem. It is stored in a chemically stable form until it becomes economically viable to enrich it for recycling.

Plutonium can be recycled and made into MOX fuels. Recycling saves enriched uranium, though the amount of money saved depends on the market price of the material, as the special precautions required for handling plutonium make it more expensive to manufacture MOX fuels than enriched-uranium fuels. On the other hand, recycling avoids the ever-increasing quantities of plutonium inside the wastes. Recycling also paves the way for effectively reclaiming the energy resources that this plutonium represents, through breeding (see chapter 12).

When it is received for reprocessing, the spent fuel has already been stored in the spent fuel pool between 6 months and a year for some of its activity to decrease (cooling). This cooling continues in one of the pools at the reprocessing plant and the fuel will not be reprocessed for at least another 3 or 4 years (in fact, the average figure is closer to 8 years). The decrease in radioactivity is quite significant in the first few years and considerably simplifies subsequent operations.

Shearing

Shearing machines are used to cut off the end-pieces from the fuel, which are then sent to be rinsed, and the fuel rods cut up into segments roughly 35 mm in length, which drop into the dissolver via a feed spout.

Dissolution

The dissolver is a wheel fitted with buckets. It rotates 30° at a time, completing a full revolution within 8 hours, thereby ensuring continuous operation. The lower part of the wheel is bathed in nitric acid. At the end of its travel, or about every 2nd hour, each bucket empties its contents through a funnel to a helical ramp where the shells are rinsed. The solution is fed off-line to a centrifuge to separate the insoluble particles from the clarified solution.

End-pieces are rinsed in acid, then in water, and go to join the shells in a storage drum ready for compacting. Compacting reduces the volume of waste to a quarter of its initial value.

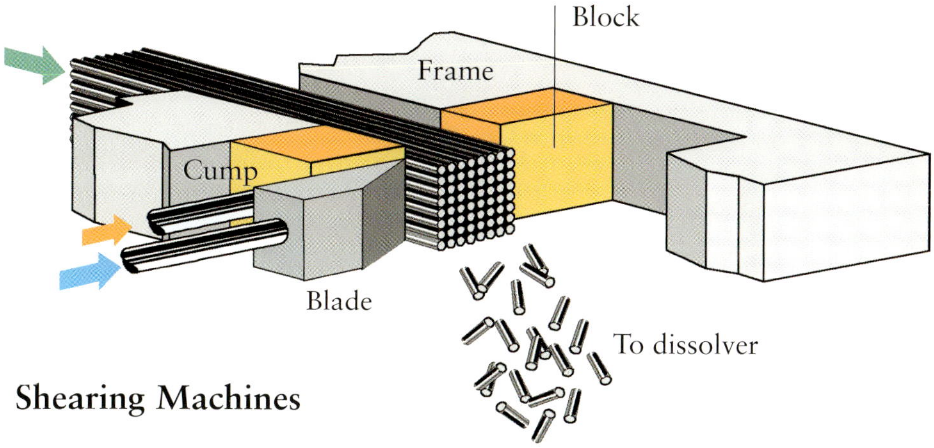

Shearing Machines

Gases released during the dissolution are thoroughly purified (dust removal and chemical treatment) before exiting through the chimney.

Extractions

Extractions are at the heart of the process as they are used to separate and purify U and Pu, which can then be recycled. The remarkable thing is that this can all be done using a single solvent, tri-butyl-phosphate (or TBP), diluted in a hydrocarbon rather similar to kerosene.

The separation stage is divided into two phases: mixing and settling. In the mixing phase, a uranium- and plutonium-rich aqueous solution of fission products is stirred with the organic phase and forms an emulsion (rather like vinegar and oil). During this phase, the uranium and plutonium show a preference for the organic phase, while the fission products and minor actinides have the opposite tendency. The second phase, settling, involves separating the "vinegar" from the "oil". This operation is repeated several times through counter-flow extraction (a classic method in chemical engineering) and the U and Pu extracted in the TBP.

The Pu and U now have to be re-extracted. The method used is to separate the plutonium and uranium right from the first cycle by selective re-extraction of the plutonium, achieved by reducing the plutonium.

It now remains to re-extract the uranium. This is done using water containing a small quantity of nitric acid. Extraction

Rotating dissolver, hanging above the dissolution vessel.

Vitrification is the best way to condition all fission products and actinides. Glass will contain almost all the fuel's radioactivity (about 99.5%)

is favored by a high nitrate content, while re-extraction is favored by a low nitrate concentration.

This highly simplified description does not take into account the multiple washing operations, solvent and effluent cycling, nor all the on-line measurements.

At the end of the first cycle, the plutonium and uranium will have been separated not only from the fission products, but also from each other. A second cycle is now carried out to obtain the degree of purity required for obtaining fuel. In chemical terms, this involves extraction with TBP in exactly the same way as before. This time, the operations are simpler as the U and Pu have already been separated from each other.

Once separated, the uranium is recovered as uranyl nitrate. Some of this is recycled as fuel after re-enrichment, while the rest is stored in oxide form, to constitute safety stockpiles.

The plutonium is transformed into its oxide through oxalic precipitation and calcination and then used to make MOX fuel.

Vitrification

Vitrification is the best way to condition all fission products and actinides. Glass will contain almost all the fuel's radioactivity (about 99.5%). This is known as ultimated waste, meaning waste that has been stabilized and which no longer contains any reusable elements.

Three different flows converge toward the vitrification facility:

- fission product solutions resulting from solvent separation
- dissolution particles separated in a centrifuge,
- effluent concentrates from the solvent treatment phases.

Each of these flows is directed to a supply preparation tank in strictly specified proportions, and additives are introduced to ensure that the glass melts properly. The liquid is constantly mixed to keep the solid particles in suspension and ensure homogeneity.

Once the load is ready, it is fed into the calciner, which is a rotating tube whose outer shell is heated to about 600 °C. The water evaporates, the nitrates are partially broken down, and the load becomes a solid calcine. It then drops into the melting furnace at the same time as a carefully-measured amount of "glass frit" granulates, composed primarily of silica, and boron oxide.

When the furnace is full, melting is continued for several hours to refine the glass. Refining here means ensuring that all the different ingredients have been oxidized and incorporated and that the glass obtained is perfectly uniform. The melting temperature must be kept above 1,050 °C. When refining is completed, the melting furnace is emptied. Each melt represents 200 kg of glass, and two melts are required to fill a 150-liter glass canister (180 liters overall volume). This glass canister contains on average 84 kg of fission products and actinides, which is equivalent to the annual electricity consumption of 400,000 families.

The end product is left to cool (in the literal sense of the term) for 24 hours, before being stored. It will not be removed from storage until it is ready to be sent to its final disposal site.

Actual vitrified waste container, on the floor of the storage building in The Hague.

Nuclear Power in the World

History	84
EPR, the Evolutionary Power Reactor	87
USA	88
France	90
Japan	92
Germany	93
Russian Federation	94
United Kingdom	96
European Union	97
South Korea	98
China	98
India	99

In 2005, 441 nuclear reactors, operating in 32 countries produced 2620 billion electric kilowatt-hours, a total of 16% of the world's electricity consumption (Table opposite) and the same amount of electricity as that generated by all the hydro-electric dams in the world. To generate the same amount of electricity from oil-fuelled plants, we would need over 600 million extra tons of oil, more than the annual output of Saudi Arabia.

History

It was back in 1951 that the small EBR1 reactor (Experimental Breeder Reactor #1) in Idaho Falls (USA), became the first reactor to produce electricity, albeit a tiny amount. The first reactor actually designed to produce electricity was connected to the grid in June 1954 at Obninsk, near Moscow and boasted an electrical output of 5 MW. The first western nuclear plant, Calder Hall, was inaugurated in 1956 by Queen Elizabeth II of England.

The fifteen following years saw nuclear reactors of many different types springing up in twenty or so different countries. There were a lot of differences between them, and some similarities:

- extraordinary creativity, especially in the United States, where almost all the possible types of reactors were tested.
- A complete lack of standardization, each reactor being one-of-a kind, with the notable exception of the submarine reactors.
- A rapid escalation of sizes and ratings, from 40 to 1100 MWe.
- The gradual introduction of laws and regulations governing the use of the new technology.
- Widespread public acceptance of nuclear power, together with strong government backing.

The mid-70s to the late 90s saw the introduction of a second generation of nuclear reactors, the ones which generate most of the nuclear power in the world today. The main features of these reactors were as follows :

- The wide range of reactor types was pruned down to a handful of reactor designs, predominantly light water reactors, PWR and BWR (see chapter 5).
- Reactor ratings stabilised at around 1000 MWe, within a range of 600 to 1400 MWe
- The implementation of a standardization policy by a number of countries, including France, bringing considerable economies of scale.
- Growing local and international public hostility, partly stemming from opposition to nuclear weapons.
- Drastic tightening of procedures and regulations.
- A serious-rated accident in March 1979 in Unit 2 of America's Three Mile Island nuclear plant (Pennsylvania) and a severe-rated accident in April 1986 in Unit 4 of the Soviet nuclear plant of Chernobyl, Ukraine (see chapter 11).

The Earth from the sky, at night, NASA composite picture showing the main energy-consuming regions.

World Nuclear Power Reactors 2005-06

	Nuclear Electricity Generation 2005		Reactors operable May 2006		Reactors under Construction May 2006		Reactors planned May 2006	
	Billion kWh	%	No	MWe	No	MWe	No	MWe
Argentina	6.4	6.9	2	935	1	692	0	0
Armenia	2.5	43	1	376	0	0	0	0
Belgium	45.3	56	7	5728	0	0	0	0
Brazil	9.9	2.5	2	1901	0	0	1	1245
Bulgarie	17.3	44	4	2722	0	0	2	1900
Canada	86.8	15	18	12595	0	0	2	1540
China	50.3	2.0	10	7587	5	4170	5	4600
Czech Rep.	23.3	31	6	3472	0	0	0	0
Finland	22.3	33	4	2676	1	1600	0	0
France	430.9	79	59	63473	0	0	1	1630
Germany	154.6	31	17	20303	0	0	0	0
Hungary	13.0	37	4	1755	0	0	0	0
India	15.7	2.8	15	2993	8	3638	0	0
Iran	0	0	0	0	1	915	2	1900
Japan	280.7	29	55	47700	1	866	12	14782
North Korea	0	0	0	0	1	950	1	950
South Korea	139.3	45	20	16840	0	0	8	9200
Lithuania	10.3	70	1	1185	0	0	0	0
Mexico	10.8	5.0	2	1310	0	0	0	0
Netherlands	3.8	3.9	1	452	0	0	0	0
Pakistan	1.9	2.8	2	425	1	300	0	0
Romania	5.1	8.6	1	655	1	655	0	0
Russia	137.3	16	31	21743	4	3600	1	925
Slovakia	16.3	56	6	2472	0	0	0	0
Slovenia	5.6	42	1	676	0	0	0	0
South Africa	12.2	5.5	2	1842	0	0	1	165
Spain	54.7	20	8	7442	0	0	0	0
Sweden	69.5	45	10	8938	0	0	0	0
Switzerland	22.1	32	5	3220	0	0	0	0
Ukraine	83.3	49	15	13168	0	0	2	1900
U. Kingdom	75.2	20	23	11852	0	0	0	0
USA	780.5	19	103	98054	1	1065	0	0
WORLD**	2626	16	441	369,374	27	21,051	38	40,737

*** The world total includes 6 Taiwanese reactors with a combined capacity of 4884 MWe, which generated a total of 38.4 billion kWh in 2005 (accounting for 20% of Taiwan's total electricity generation). Taiwan has two reactors under construction with a combined capacity of 2600 MWe.*

Operating : 441 Units 381 GWe

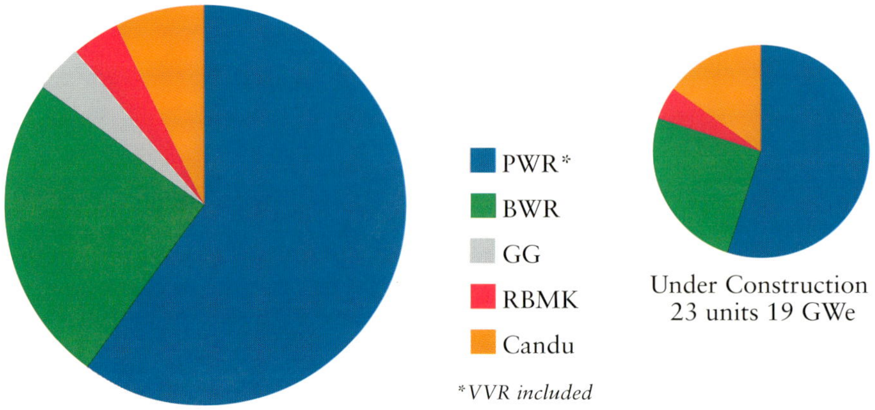

GWe Installed in the World

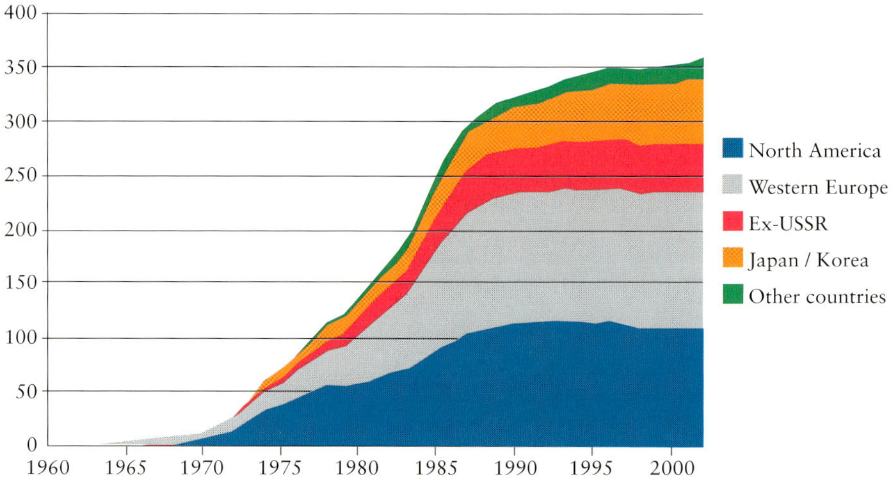

The major reactor manufacturers now supply 3rd-generation reactors, which were designed after the Chernobyl accident and include safeguards to prevent any massive release of radioactivity into the environment in the event of a severe accident, including full core meltdown. Generation III reactors have also been given excellent protection against external attack.

Typical of the 3rd generation reactors is AREVA's EPR, Evolutionary Power Reactor. A prototype, with a rating of 1600 MWe, is under construction on the Olkiluoto site in Finland (see opposite page). EPR's competitors include Westinghouse's AP 1000 reactor and General Electric's ESBWR reactors, although none have yet to be ordered (2007).

TVO

Generation III reactors should answer the need for nuclear plants over the next few decades. Looking further into the future, work has already begun internationally on 4th-generation reactor projects (see ch.12), which will in all likelihood offer the following features

- Ready for commercialization around 2040 ;
- More efficient use of fissile materials (see chapter 12) ;
- Producing less long-lived radioactive waste;
- Designed to enlarge the scope of applications of nuclear energy beyond electric power alone (fresh water, process heat, hydrogen production, etc.).

EPR, Evolutionary Power Reactor

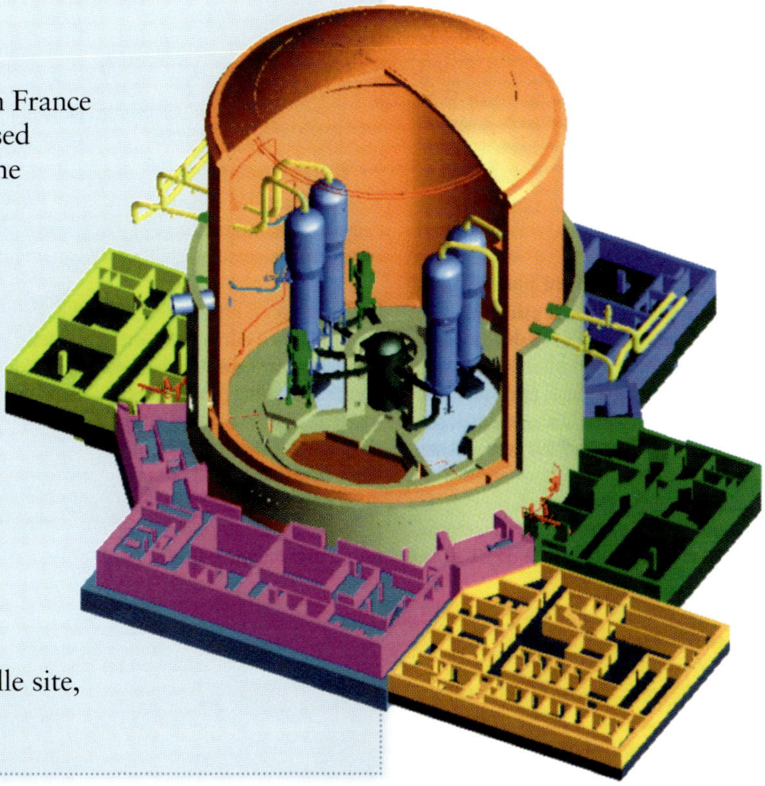

Derived from the most modern PWR operating in France (N4 series) and Germany (Convoy series) and based upon an unrivalled return of experience, EPR is the archetype of Generation III reactors and features safety systems (4 redundant trains located in separated buildings) that ensure safe operation. Even in the unlikely event of a severe accident, special design-features would prevent the dissemination of radioactivity outside the reactor building. Additional reinforced concrete shells protect key buildings from terrorist attacks or accidental external damage. Despite the cost of these safety features, EPR is economically competitive due to its improved availability, higher thermal efficiency and longer design lifetime. After the Olkiluoto reactor in Finland, the French are preparing to build a lead plant on the Flamanville site, in northern France.

EPR site in Olkiluoto, Finland: the future EPR is on the left of the two current reactors.

United States

In 2005, the USA had 103 nuclear reactors operating on 65 sites in the country, with a total capacity of 98 GWe, generating 780 billion kWh, or 19% of total US electricity production. The USA is therefore the largest nuclear power producer in the world and nuclear energy is the second largest source of US electricity, although far behind coal. Their nuclear fleet is made up exclusively of light water reactors: 2/3 PWRs and 1/3 BWRs. Average reactor availability has been exceeding 90% for several years.

President George W Bush visiting the control room of the Calvert Cliffs nuclear plant, near Baltimore, Maryland, in 2005.

Nuclear development in the United States has had contrasting fortunes: it started on a fast track in the 60s, and then experienced a brutal stop in 1974, following the first oil crisis – the very same oil crisis which triggered the spectacular acceleration of the French nuclear program. The accident which occurred in March 1979 in unit 2 of the Three Mile Island (Pennsylvania) nuclear plant did a lot to prevent a revival of the US program, and the industry had still not recovered in 2006.

Although the reactors in operation or under construction in the USA were all ordered before 1974, the latest reactor on line was only connected in 1997. On the other hand, over the last 12 years, while the total number of reactors licensed to operate has gone slightly down, nuclear electricity generation in the United States has demonstrated the same growth as if some twenty new reactors had been put on line!

The two main weaknesses of the US program have been the near total lack of standardization and the scattering of the nuclear power fleet: When there were 108 operating reactors, they belonged to 63 different electric utility companies. Over the last decade, a handful of companies (Constellation, Dominion, Entergy, etc.) have bought second-hand reactors from utilities which owned only one or two units and have built

Nuclear Plants operating in the United States

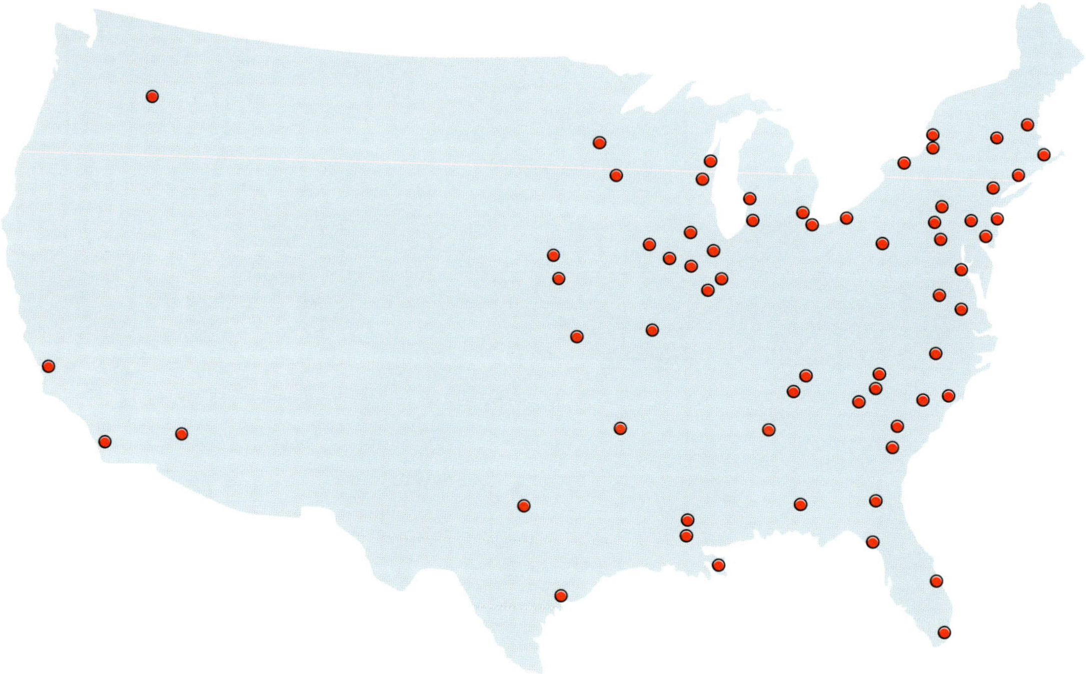

The USA is the world's largest producer of nuclear power, which constitutes the second biggest source of electricity, far behind coal.

up fleets which can be managed more efficiently and more economically.

In the Energy Policy Act, enacted in July 2005, the US Congress voted a package of measures to provide financial help for the companies ordering the first six new nuclear reactors. These measures, together with the licensing reform (Design Certification and Early Site Permit) and the spectacular improvement in operating reactor performance over the last decade have caused a resurgence in demand over the next few years, with the first new orders coming in since 1974. Three companies are offering generation III reactors: Westinghouse's AP 1000, General Electric's ESBWR and AREVA's EPR.

The USA used to have four major reactor vendors, but only two remain in business: General Electric, the designer of the BWR, and Westinghouse (since 2006 an affiliate of Japan's Toshiba), the first manufacturer of the PWR. Uranium has been extensively mined in the USA and only two mines currently remain in operation. Uranium needs are met by imports, mainly from Canada, but also from Australia, Kazakhstan and Uzbekistan.

Uranium enrichment operations were transferred in 1993 from the Department of Energy to the US Enrichment Corporation, a private company. USEC operates an ageing gaseous diffusion plant on the Paducah site belonging to DOE and commercializes the low enriched uranium obtained from the down-blending of 500 tons of highly enriched uranium which used to be part of Soviet military stockpiles. US utilities import enrichment services from Europe (Eurodif, Urenco) and Russia. The USA will soon be switching to ultracentrifuge technology (chapter 6).

Three companies manufacture nuclear fuel in the United States: General Electric, Westinghouse and AREVA.

Even though America invented spent-fuel reprocessing technology, as part of its Defense programs, their civilian reprocessing industry was killed off in its infancy when Gerald Ford, during his ill-fated re-election campaign against Jimmy Carter, banned civilian reprocessing in 1975, as part of a non-proliferation policy that had not yet found its bearings. The veto was subsequently lifted by Ronald Reagan, but the industry was by then no longer willing to invest in the technology.

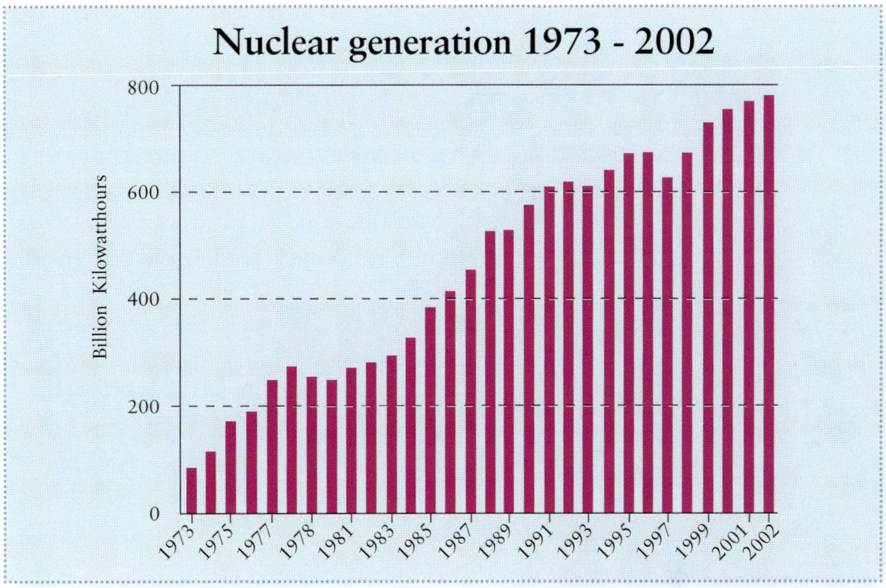

The Nuclear Waste Policy Act, enacted in 1982, made the government responsible for safely disposing of spent fuel from the electric utility companies, which, with the ban on reprocessing, had built up as high-level radioactive waste. The Act committed the Department of Energy (DOE) to taking over the spent fuel by January 1998 at the latest. It also established a 1.0 mill ($0.001) per kilowatt hour tax on nuclear power, to be deposited in the Nuclear Waste Fund and be used for financing a repository. DOE selected the site of Yucca Mountain, in the Nevada desert, to dispose of spent fuel inside a volcanic ridge-line. Strong opposition from the State of Nevada and a number of adverse features in the site have made it difficult to forecast the date when the disposal site will be opened. It is not even completely clear whether DOE will use Yucca Mountain as a repository or as a long term storage site.

In 1999, DOE launched the "Generation IV" initiative, which set out the main framework for international R&D on future nuclear systems (see chapter 12).

US Senator Pete Domenici, the main architect of the 2005 Energy Policy Act.

The USA is a party to the Non Proliferation Treaty NPT as a Nuclear Weapon State NWS

Joseph Immelt, CEO of General Electric.

AREVA offers a comprehensive range of nuclear services for operators across the world

France

In 2005, France had 59 nuclear reactors operating on 21 sites in the country, with a total capacity of 63 GWe, generating 431 billion kWh, or 79% of total French electricity production. Nuclear power is thus the main source of French electricity, far above hydropower. France ranks in the world number 2 for nuclear power generation, and number one in terms of percentage of nuclear electricity. Given their fairly unique operating environment, French nuclear reactors do not solely operate "baseload", and must to a certain extent accommodate load following*. Fleet availability factor runs at around 84%.

French Nuclear Plant in Cattenom (Moselle).

With the exception of the 250 MWe Phénix experimental FBR, French nuclear power plants are all PWRs and show a very high degree of standardization, based on three types: thirty-four 900 MWe units, twenty 1300 MWe units and four 1450 MWe units.

The French nuclear power program started back in the 60s and was based on a nationally- developed reactor design, the UNGG, which was graphite-moderated, cooled by pressurized carbon dioxide and fuelled by natural (unenriched) uranium. In 1970, the EDF (Electricité de France) national utility decided to switch to PWR and BWR light water reactor technology, built under US licence at a rate of 2 units per year. By then, the CEA (the French atomic energy commission) had mastered the gaseous diffusion enrichment technology, which made the decision politically that much easier.

**Load following is when an electricity utility adds extra generation to available electricity supplies to meet fluctuating demand, and/or keeps generating facilities informed of load requirements to ensure that generators are producing the right amount of energy to supply the utility's customers.*

In 1974 came the first oil crisis and the French government reacted by authorizing EDF to drastically step up its nuclear program and order 6 units per year. The following year, EDF decided to base its nuclear plants exclusively on PWRs, and gave the contract for building them to Framatome (now AREVA NP). Westinghouse was requested to sell its stake in Framatome to the CEA and in 1976, the CEA transferred its fuel cycle businesses (mines, enrichment, reprocessing) to COGEMA, a fully owned industrial company (now AREVA NC). This phase ended in 2000 when unit 2 of the Civaux nuclear plant was put into industrial operation. Renewal of the French nuclear fleet should begin around 2020 and EDF started gearing up for this in 2006 with an order for a Generation III EPR lead plant (see page 87), to be built on the Flamanville site, in Normandy.

The French nuclear industry today has a fairly straightforward structure. EDF is the country's sole nuclear power plant operator, while AREVA offers a full range of nuclear services which it also sells to nuclear operators worldwide. The CEA performs a large part of the R&D needed to support civilian nuclear activities and ANDRA, the national radioactive waste management agency, is responsible for the disposal of ultimate waste (paid for by the waste producers).

France has used up most of its domestic uranium resources, and the last French uranium mine was decommissioned in 2001. AREVA NC is still, however, the world's second largest uranium producer, behind Canada's CAMECO and exploits uranium mines in several countries including Canada and Niger. AREVA is also the main shareholder of EURODIF, the European Gaseous Diffusion Uranium Enrichment Consortium. In 2012, a new ultracentrifuge enrichment plant is due to replace the existing EURODIF Georges Besse plant (chapter 6).

AREVA manufactures nuclear fuel from both enriched uranium and plutonium (MOX), for PWR and BWR reactors,

Anne Lauvergeon, CEO of AREVA.

Nuclear Plants operating in France

while AREVA NC operates the world's largest spent-fuel reprocessing plant in The Hague, for PWR and BWR. The Hague also conditions the radioactive waste resulting from the reprocessing.

France has opted to reprocess its spent fuel so as to recycle the residual fissile materials and optimize conditioning of the ultimate waste: 20 EDF PWRs routinely recycle the plutonium recovered by spent fuel reprocessing as MOX fuel assemblies, while 2 units are fuelled with re-enriched recovered uranium.

Until 1997, France was world leader in the development of fast breeder reactors. The shutdown of Superphoenix, a prototype FBR with a power output of 1200 MWe, proved to be a major blow, but despite this setback, France was the first country to join the United States in 1999 as part of the Generation IV initiative, in which the FBR design is a prominent feature.

Low-level and very low-level radioactive waste are disposed of in the surface storage sites of Soulaines and Morvilliers (Aube). A law enacted in June 2006 set out the roadmap for ultimate management of long-lived radioactive waste, while an underground laboratory operated by ANDRA in eastern France is characterizing a clay stratum where a deep repository for long-lived radioactive waste could be built by around 2020. Research is also being carried out to look into the feasibility of transmuting the long-lived fraction of the radioactive waste.

France is a party to the Non-Proliferation Treaty, NPT, as a Nuclear-Weapon State NWS

Japan is a party to the NPT (Non-Proliferation Treaty) as a Non-Nuclear-Weapon State NNWS

Japan

In 2005, Japan had 55 nuclear reactors operating on 16 sites in the country, with a total capacity of 48 GWe, generating 280 billion kWh, or 29% of total Japanese electricity production. Nuclear energy is the second biggest source of electricity in Japan, behind coal, and Japan is the third largest nuclear power producer in the world.

Apart from the 300 MWe prototype FBR Monju, all Japanese plants are light water reactors and the fleet comprises 30 BWRs and 24 PWRs. The reactors are owned and operated by the 10 main Japanese utilities, including TEPCO, which supplies electricity to Tokyo and Kansai, which serves Osaka. The PWRs, initially constructed under Westinghouse license, are built by Mitsubishi Heavy Industries while the BWRs, initially constructed under General Electric license, are built by Hitachi and Toshiba, which explains why Toshiba's acquisition of Westinghouse came as such a surprise and resulted in tighter links between GE and Hitachi.

The Japanese nuclear programme has been growing steadily for some thirty years without any abrupt halt like in the US or steep acceleration like in France. The Japanese public has not forgotten the traumas of Hiroshima and Nagasaki, but it is acutely conscious of the country's lack of natural energy resources, especially oil. Japan's dependence upon energy imports is 81%, but it would reach 96% without nuclear power. All the nuclear power plants are located by the sea, and the licence procedure of every site has been subject to lengthy and intense negotiations with the local fishermen's associations.

Japan imports uranium mainly from Canada, Australia and Niger. Japan Nuclear Fuel Ltd operates a centrifuge enrichment plant but most of the enrichment services are imported from the USA (USEC) and Europe (EURODIF, Urenco). Fuel manufacture is partly domestic and partly imported.

Japanese utilities used to have most of their spent fuel reprocessed in Europe (BNFL and AREVA), while JNC operated a small plant in Tokai Mura. In 2006, a large reprocessing plant (800 t/year) began operating in Rokkasho Mura. The Pluthermal program for recycling plutonium in light water reactors is due to start, but the Japanese are firmly committed to the long-term development of fast neutron breeders. The prototype FBR Monju has been shut down since 1995, following a sodium leak, but repairs were given the go-ahead in 2006 and are currently under way.

An experimental 30 MWt HTR (high temperature reactor), operating on the Oarai site, will be the first to demonstrate on a pilot scale the production of hydrogen from nuclear power.

Japan is a party to the NPT (Non-Proliferation Treaty) as a Non-Nuclear-Weapon State NNWS.

Nuclear Plants in Japan

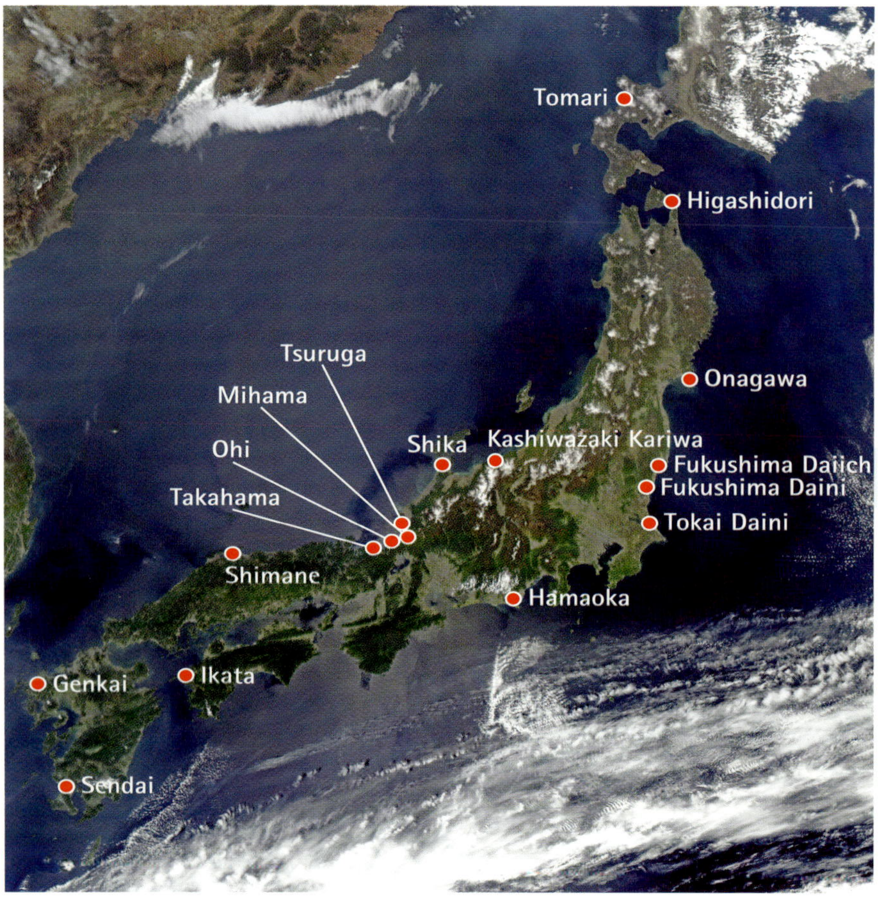

Germany

In 2005, Germany had 17 nuclear reactors operating on 12 sites in the country, with a total capacity of 28 GWe, generating 155 billion kWh, or 31% of total German electricity production. Germany is the 4th largest world producer of nuclear power, which is its second biggest source of electricity, behind coal (and lignite).

However, in 2001 Germany decided to phase out nuclear power and, if the current law is not amended, the last German reactor will be decommissioned in 2021. Most of the German nuclear program was implemented during the 70s, at the rate of 3 units per year, while the anti-nuclear opposition was becoming stronger and resorting to increasingly violent methods.

All currently-operating reactors (11 PWRs and 6 BWRs) are located in the former West Germany as all the reactors which used to operate in the eastern part (VVRs of Soviet design) were shut down following unification.

The two largest German reactor manufacturers, Siemens and AEG, merged their nuclear power divisions in 1989 and in 1999, the business was transferred to Framatome ANP (now AREVA NP), in which Siemens has a 34% shareholding. The joint venture meant the experience and know-how of German nuclear scientists and engineers (who had already helped design the EPR), could be kept, despite the German phaseout.

There is no uranium mine operating in Germany (the once large reserves of the former East Germany were completely exploited during the Soviet era). German power producers E.ON and RWE are partners in the Anglo-Dutch-German centrifuge enrichment consortium, Urenco

German utilities used to have their spent fuel reprocessed, notably in the The Hague UP3 plant, and the recovered plutonium was recycled as MOX fuel in German reactors. However, spent fuel transportation was discontinued in 2005, in line with the nuclear phaseout law.

Nuclear Plants in Germany

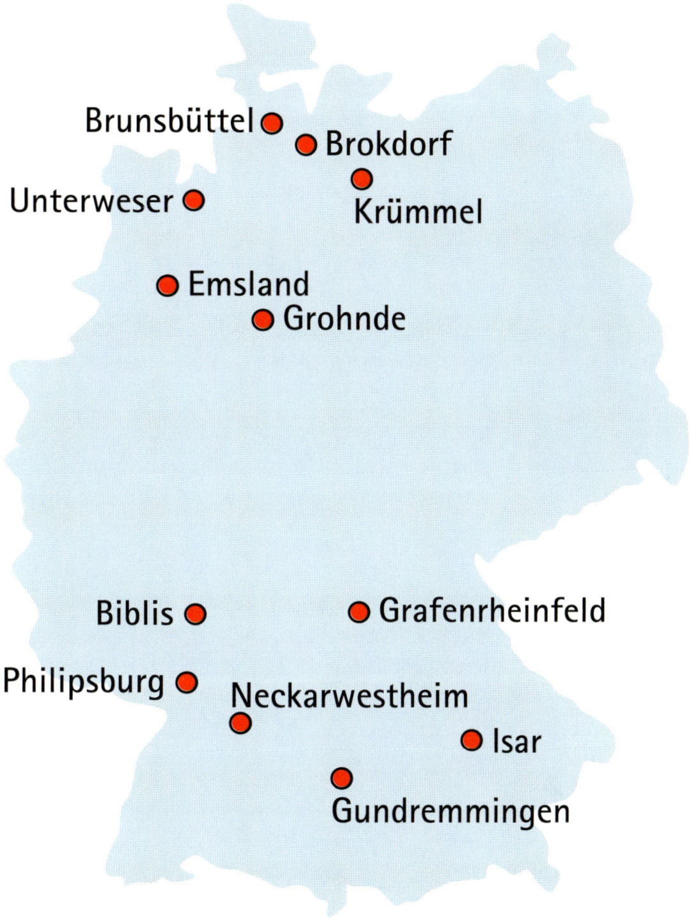

Anti-nuclear forces have been particularly vociferous in their opposition to the high-level radioactive waste repository planned on the Gorleben site.

During the 70s and 80s, the German R&D centres of Jülich and Karlsruhe were actively involved in the preparation of future reactor designs. Germany built a 300 MWe HTR prototype as well as a prototype FBR of similar rating, while German utilities were a partner in the French-built Superphoenix. All R&D on future nuclear designs, except fusion, has now been banned in Germany and the country is therefore not part of the Generation IV imitative. The German "pebble bed" HTR technology has been transferred to South Africa and China (see ch.12).

Germany is a party to the NPT (Non-Proliferation Treaty) as a Non-Nuclear- Weapon State NNWS.

Angela Merkel, when President of the CDU, advocated abrogating the nuclear phaseout law. She became German Chancellor in 2005.

In 2006, Russia announced the launch of an important nuclear power program, aimed at commissioning 2 GWes per year, starting in 2011, and at providing 25 to 30% of its electricity from nuclear energy by 2030.

The Russian Federation

In 2005, the Russian Federation had 31 nuclear reactors operating on 11 sites in the country, with total capacity of 22 GWe, generating 137 billion kWh, or 16% of Russia's total electricity production (same percentage as the world total). The Russian Federation is the 6th largest world producer of nuclear power.

The Soviet nuclear program was going full out between 1964 and 1989, at which date 10 units were under construction. The collapse of the Soviet Union put the nuclear program into hibernation for ten years or so, but it started up again in 2001 with the commissioning of a new plant in Rostov.

Apart from a 600 MWe FBR in Beloyarsk and 4 small dual-purpose (district heating and electricity) units in Bilibino, Russian reactors belong to two different families: the 440 and 1000 MWe VVRs are eastern versions of PWR while the 1000 MWe RBMKs are graphite-moderated and cooled by boiling water (Chernobyl type). The RBMKs, in service since 1973, were used both to generate electricity and to produce weapon-grade plutonium for the Soviet arsenal. This explains why they were never sold to the "satellite" countries, but were built solely in the USSR for use in Russia, Ukraine and also Lithuania, now a member of the European Union.

All the Russian nuclear plants are operated by ROSENERGOATOM.

The Russian Federation has been actively exporting nuclear reactors and 5 units were under construction in 2005, with 2 in China, 2 in India (where Russia broke the embargo which followed India's nuclear explosion in 1974 and its subsequent refusal to join the Non-Proliferation Treaty), and 1 in Iran, which is the focus of an escalating dispute with the USA.

It is mainly in the fuel cycle that Russia is a major player. Their huge facilities, initially built for Defence purposes, were dimensioned in the 80s for a civilian program of around 100 GWe, which gave the country a huge export potential.

The Russian Federation has produced 120,000 metric tons of uranium since 1944. The sheer size of the country implies that its ultimate resources are still sizeable: Reasonably Assured Reserves alone are close to 130,000 tons (cost < $80/kg U), while annual production amounts to roughly 3,200 t/year.

4 centrifuge enrichment plants have a total output of 15,000 SWU/an, and the capacity of the two fuel fabrication plants is 2,500 t/y, enough to supply all global RBMK and VVR reactor needs.

Russia operates a spent-fuel reprocessing plant of 400 t/y nominal output, but its actual production does not exceed 150 t/y. The recovered uranium is recycled in RBMK units, while plutonium is stockpiled to fuel future FBR reactors. As part of US-Russian disarmament agreements, a program demilitarizing weapon-grade plutonium by burning MOX fuel in power reactors is under way, albeit at a slow pace.

The Federal Atomic Energy Agency, established in March 2004 as the successor to MINATOM, is in charge of all fuel cycle activities. A huge organization, the FAEA has 450,000 employees. A presidential Bill, approved by the Duma in January 2007, set up a state-controlled holding company, Atomenergoprom, using the sector's civilian assets. The Bill aims to allow other Russian enterprises to possess non-weapons-grade nuclear materials, nuclear installations and nuclear storage facilities.

The Soviet Union used systematically to take back spent fuel from its satellites (and Finland), for the same reason it allowed export only of VVR plants. This practice, similar in many ways to leasing, is again being considered in the Russian Federation, although there is some ambiguity surrounding the fate of the ultimate waste.

Russia injects part of its own high level waste directly, in liquid form, into a geological stratum, while the rest is stored in surface storage facilities.

High-level nuclear research is carried out in a number of research institutes

Nuclear Plants in the Russian Federation

(Kurchatov Institute, NIKIET and VNIM in Moscow, IPPE in Obninsk, NIIAR in Dimitrovgrad, etc.).

Since Superphoenix was decommissioned, the Beloyarsk reactor is now the largest currently-operating fast neutron reactor (it is not a breeder as, until now, it has been fuelled with enriched uranium instead of plutonium). The fast neutron reactor BN350, located in Aktau, Kazakhstan, and decommissioned in 1999, was the first nuclear reactor to produce vast amounts of potable water (chapter 8), while the smaller Bilibino units are also the largest nuclear district heating facilities (Siberia is a highly logical choice for this kind of application). The Tokomak nuclear fusion reactor, described in chapter 13, was also invented in Russia.

To prevent a large-scale brain-drain following the collapse of the Soviet Union, an international program, managed by the International Science and Technology Center ISTC, provides direct funding for civilian R&D programs in Institutes formerly given over to nuclear weapons development. Many military nuclear facilities, another legacy of the Soviet era, are derelict and demand costly and lengthy dismantling and cleaning operations. The decommissioned nuclear submarines are probably the best known examples.

The Russian Federation is a party to the Non-Proliferation Treaty NPT as a Nuclear-Weapon State NWS.

Engineers working on unit 3 of the Kalinin nuclear plant (near Tver, Russian Federation).

United Kingdom

Queen Elizabeth II inaugurated the first western nuclear power plant at Calder Hall, in 1956.

In 2005, the UK had 23 nuclear reactors operating in the country (England, Scotland and Wales), with a total capacity of 12 GWe, generating 75 billion kWh, or 20% of Britain's total electric production. Nuclear power is the UK's second largest electricity source behind gas, in a country without any sizeable mountains to speak of and therefore not much hydropower potential. The Thatcher years also saw a massive switch from coal to gas (as a result, the UK will meet its commitments under the Kyoto Protocol).

The UK was the first western country to make huge investments in nuclear power, with the Magnox series and the AGR, both of which are graphite-moderated and cooled by pressurized carbon dioxide. The discovery of oil and gas deposits in the North Sea proved to be a severe blow for the then-flourishing nuclear industry. There is only one modern reactor in operation, the Westinghouse-designed 1200 MWe Sizewell B PWR, commissioned in 1995. If the current decommissioning rate keeps up, and barring new orders, by 2020, Sizewell B will be the last remaining nuclear reactor in operation.

The UK was the first western country to massively develop the generation of nuclear power

The UK used to export natural gas, but has been a net importer since 2004, and is now having to face the steep escalation in gas prices in the international market, just when gas has become the country's largest source of electricity, and the decommissioning of older nuclear plants has been speeded up. The situation led Tony Blair's government to seriously reconsider the role of nuclear power in the country's energy supply (a role completely ignored in the 2003 White Paper). The present trend seems to be leading to a renewal of the nuclear fleet to bring it up to a certain level.

The Calder Hall nuclear power plant.

Until 1990, electricity generation, transportation and distribution were split between two utilities: CEGB in England and Wales and SSEB in Scotland. In 1990, the system was deregulated and a new private company, British Energy (BE), set up to run the AGR and PWR nuclear plants, while the older Magnox facilities were transferred to the state-owned BNFL (British National Fuels Ltd). However, shorn of the lucrative transportation and distribution business, BE found itself losing a lot of money as low gas prices were driving electricity prices down.

On April 1, 2005 the government set up the NDA, the Nuclear Decommissioning Authority,, a public body tasked with taking over from BNFL (with the agreement of the Brussels Commission) the costs and commitments involved in decommissioning nuclear facilities, a financial burden which had almost driven BNFL into bankruptcy. In return, BNFL was forced by the government to sell the Westinghouse Electric Company, bought from CBS in 1999. BNFL did, however, keep responsibility for operating the Magnox plants.

Although the United Kingdom has practically no uranium of its own, BNFL is a major international player in the other sectors of the fuel cycle. The company is Urenco's British shareholder for uranium enrichment, it manufactures fuel for the Magnox, AGR and LWR reactors, and it operates the two Sellafield spent-fuel reprocessing plants. The more modern of the two, THORP, handles the reprocessing of foreign BWR and PWR spent-fuel assemblies, but was temporarily shut down in 2006 following internal radioactive leakage.

Low-level radioactive waste is stored in Driggs and Dounreay, while most of the high-level waste is in interim storage at Sellafield, awaiting the decision as to its ultimate disposal, in all probability in deep geological storage.

The United Kingdom used to be very active in FBR development, with two units on the Dounreay site. Both plants have now been decommissioned, and there is little research in the field, despite the UK being a partner in the Generation IV initiative.

The United Kingdom is a party to the Non-Proliferation Treaty NPT as a Nuclear-Weapon State NWS.

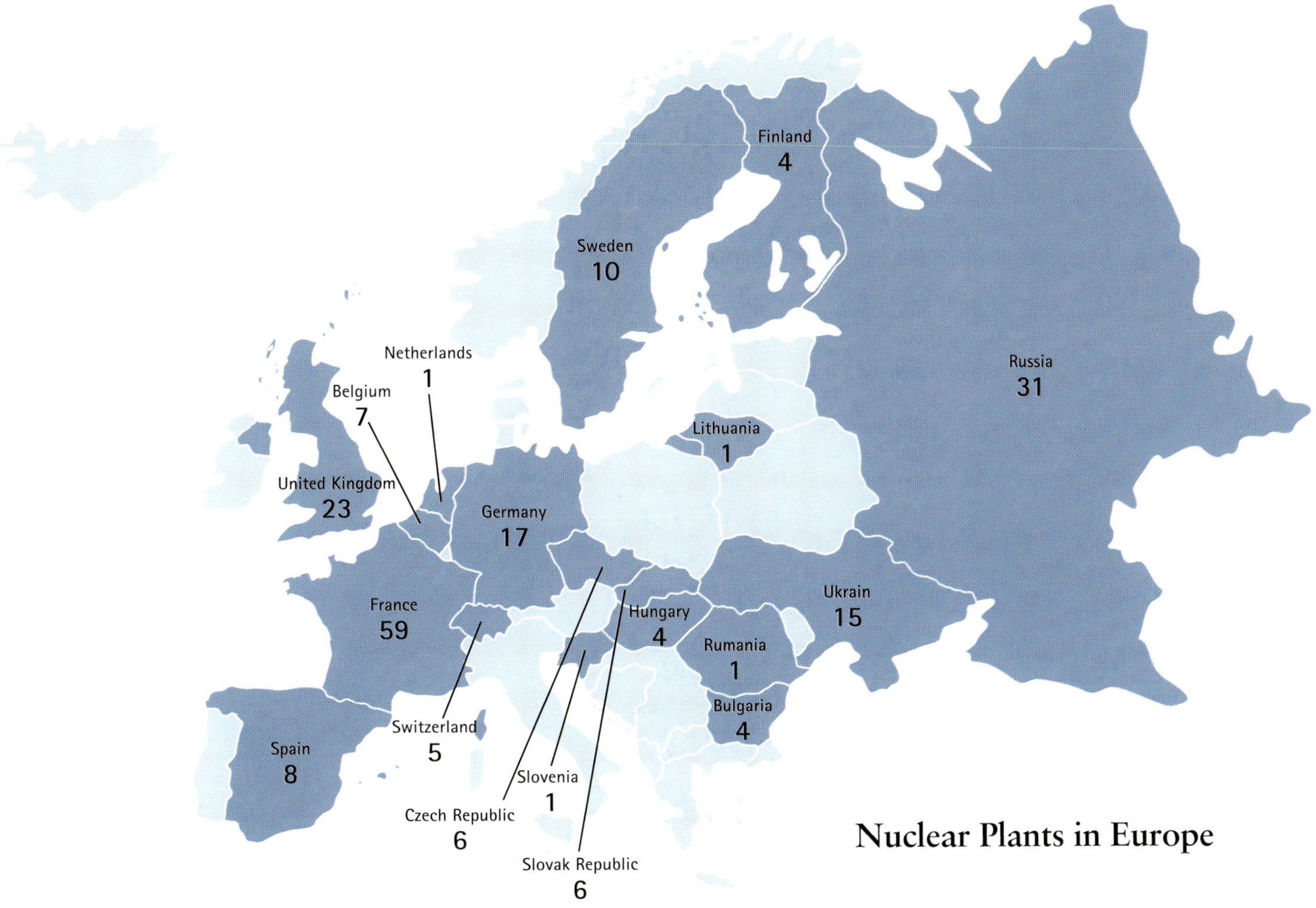

Nuclear Plants in Europe

The European Union

In 2005, there were 147 nuclear reactors operating in the 25 countries of the European Union, with a total capacity of 130 GWe, generating 925 billion kWh, or one third of total EU electricity production. Nuclear power is the largest source of electricity in the Union. The last two States to join the Union, Bulgaria and Romania, are also nuclear power producers (23 TWh in 2005).

The situation as far as nuclear power in EU is concerned is somewhat paradoxical. It is the largest nuclear-electricity-producing area in the world, both in absolute and in percentage terms, but it is also the region where there is the strongest opposition to all things nuclear. The situation varies strongly from one State to another:

Austria, Cyprus, Denmark, Estonia, Greece, Ireland, Italy, Latvia, Luxemburg, Malta, Poland and Portugal do not produce any nuclear power. Germany, Belgium, Spain and Sweden have decided, more or less formally, to phase out nuclear power in the long run. The Netherlands, Slovenia and the United Kingdom have not announced any firm plans. Lithuania, obliged by the EU to shut down its last reactor (RBMK) by 2009, wants to build a replacement nuclear plant (it might be joined by Latvia, Estonia and Poland). Bulgaria, Finland, France, Hungary, the Czech Republic, Romania and Slovakia are firmly committed to pursuing their nuclear programs.

EU Member States are also parties to the EURATOM Treaty. The Euratom part of the various Framework Programs of the EU incorporates a sizeable slice of European nuclear R&D and it provides funding for about half of the Member States, R&D programs involving nuclear fusion, the bulk of the various programs carried out by the Joint Research Center of the EU and a tiny fraction of the program involving nuclear fission.

EURATOM ASSOCIATION HAS

Europe is the largest nuclear electricity producer in the world, in both absolute and percentage terms. It is also, however, the region where there is the strongest opposition to all things nuclear

Promotion of nuclear power in South Korea.

South Korea

In 2005, South Korea had 20 nuclear reactors operating in the country, with a total capacity of 17 GWe, generating 139 billion kWh, or 45% of total South Korean electricity production. South Korea is now the 5th largest world electricity producer, just in front of Russia. Nuclear power is the country's largest source of electricity. Installed capacity could reach 40 GWe in 2020.

The Korean nuclear program is a model in terms of consistency and continuity. After a first phase in which they were buying turnkey plants, the Koreans demanded and obtained full technology transfer, with increasing localization of component production and plant construction under license. They are now fully autonomous.

Sixteen of Korea's reactors are PWRs, and the other 4 Candus. The Korean KSNP (Korean Standard Nuclear Plant) is a 1000 MWe PWR using technology derived from Combustion Engineering. The first plants of the next generation, AP 1400, are under construction.

KEPCO, the former national utility, was divided into 6 large divisions in 2001, but any idea of privatization has since been abandoned. Its affiliate, KHNP (Korean Hydro and Nuclear Plants), operates all the country's nuclear plants and is also their main architect-engineer. Another affiliate, KNFC, manufactures the fuel and has a capacity of 550 t/y of PWR fuel and 700 t/y of CANDU fuel.

South Korea has no access to uranium enrichment or spent fuel reprocessing technologies (the USA is opposed to the idea) nor does it have uranium reserves or any domestic uranium production.

KAERI has an active, ongoing R&D program, and its work includes developing the DUPIC process to burn spent PWR fuel in Candu reactors (exploiting the superior moderating power of heavy water). The Koreans have also developed a small nuclear reactor design, the SMART, for both electricity production and seawater desalination.

South Korea is a party to the NPT (Non-Proliferation Treaty) as a Non-Nuclear-Weapon State NNWS.

China

In 2005, China had 10 nuclear reactors operating in the country, with a total capacity of 8 GWe, generating 50 billion kWh, or 2% of China's total electricity production. Nuclear energy is not currently a major source of electricity, but the Chinese government is planning on expanding it to provide 4% of the country's total electricity production by 2020.

While China is one of the 5 Nuclear-Weapon States, its civilian nuclear power program is relatively recent. Development of the first Chinese power reactor, a 310 MWe PWR, started in 1981 and it was commissioned in 1991. Four of its operating reactors are French-designed PWRs, two are Russian-supplied VVRs, two are Canadian-built Candus and three are Chinese-designed. China plans to complete its program with a standardized series, and gave the nod for four AP 1000s to Westinghouse at the end of 2006. The Chinese nuclear sector is now able to manufacture most of the plant components as well as the fuel assemblies.

Uranium Reasonably Assured Resources in China are estimated at 85,000 tons, but the country is so huge that there are certainly sizeable undiscovered resources. Production capacity stands at around 700 t/y. There is no civilian enrichment or spent-fuel reprocessing.

In 2000, the Chinese commissioned a small 10 MW pebble-bed HTR at Tsing Hua University, and a 25 MWe experimental FBR is under construction in cooperation with the Russian Federation. CNNC is also working on the CNP-1000 PWR, strongly inspired by Westinghouse's AP 1000.

China is a party to the Non-Proliferation Treaty NPT as a Nuclear-Weapon State NWS.

India

In 2005, India had 15 nuclear reactors operating in the country, with a total capacity of 3 GWe, generating 16 billion kWh, or 2.8% of India's total electricity production. Nuclear power is still a very small source of electricity for the subcontinent, but its highly-ambitious program plans to have 20 GWe in operation by 2020.

India's decision to develop an independent nuclear program was not entirely by choice: as part of the Atoms for Peace program, Canada supplied India with the CIRUS 40MWt experimental heavy water reactor, a duplicate of their NRX, while the USA supplied the heavy water. India manufactured its fuel itself. CIRUS started operation in 1960, and India began to extract plutonium from its spent fuel in 1964. The plutonium was used on May 18, 1974 to carry out the famous Indian «peaceful nuclear explosion», code-named "Smiling Buddha". The Non Proliferation Treaty, drafted in 1968 and in force since 1970, only recognizes 5 Nuclear Weapons States: USA, USSR, UK, France and China. The Indian explosion caused enormous concern throughout the world, especially in the United States and Canada who had unwittingly made it possible. It was mainly as a result of India's new weapon that the USA abandoned the reprocessing of civilian spent fuel and the main nuclear technology holders created the "London Club", now known as the NSG, the Nuclear Suppliers Group, to voluntarily curb their sensitive exports.

The immediate consequence saw India - a non-signatory of the NPT - cut off from international assistance in the nuclear field until 2001 when Russia broke the taboo by selling them two VVR 1000 reactors. The USA and France look likely to follow Russia's example.

India's first two power reactors, 2 BWRs located at Tarapur, were supplied by the USA. Canada later helped India build a first Candu on the Rajasthan site. After Canada's withdrawal in 1974, the Candu was the lead plant for the whole Indian nuclear power fleet of twelve 200 MWe Candu units. The technology to produce heavy water was first acquired from France. All the power plants were designed and built and are operated by the state-owned company NPCIL, which sells bulk power to the various State Electricity Boards.

Indian uranium reserves are estimated at 85,000 tons. Annual production is close to 250 tons but capacity is increasing and should soon exceed 500 t/y. India enjoys huge thorium resources, and an advanced Candu is being designed to exploit them.

India has two important nuclear research centers (both military and civilian): the Bhabba Atomic Research Center (BARC) near Bombay and the Indira Gandhi Center, at Kalpakkam, near Madras. In the latter, a small experimental FBR, the FBTR, is operating and is based on French technology but using Indian-manufactured fuel, including a thorium blanket. Nearby, a prototype 500 MWe FBR is under construction.

With 8 reactors under construction, the Indian nuclear program is very active. It is also highly ambitious, and aims to generate 10% of the country's electricity by around 2020 and 55% by 2050 !

India is a fast-emerging country, whose economic vitality and demographic growth have caused electricity consumption to soar. If coal were to remain India's almost sole source of electricity, the climatic consequences would be catastrophic. The whole international community is therefore monitoring the development of nuclear power in India closely, but there still remains the very tricky problem of the country's stance with regard to the Non-Proliferation treaty: India is not a party to the NPT and is not recognized as a Nuclear-Weapon State, but it acquired A-bombs in 1974 and H-bombs in 1998...

Homi Bhabha, the father of India's nuclear program.

It was in 1974, during Indira Ghandi's government, that India performed its famous "peaceful nuclear explosion", code-name Smiling Buddha.

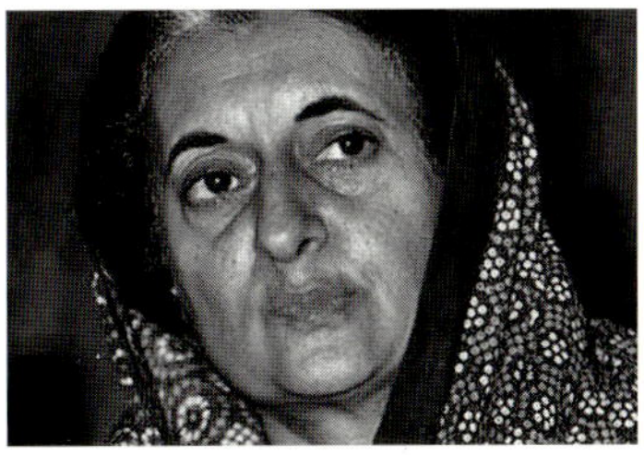

Nuclear Energy: the Economic Aspects

Capital Costs	103
Fuel Costs	107
Other Operating Costs	108
Insurance Costs	110
Waste-Disposal Costs	110
Other External Costs	111
Decommissioning and Dismantling Costs	112
Conclusion	112

Economists consider that all the methods that lead to a reduction in CO_2 emissions (taxes, tax incentives for improving energy efficiency, emission caps, emission-rights trading) boil down to a simple question of what is the cost to be allocated to a ton of carbon in the atmosphere or what is the cost of emitting a ton of CO_2 (a ton of carbon is the equivalent of 3.66 tons of CO_2).

The cost price of a kilowatt-hour of nuclear power has given rise to much argument over the last thirty or forty years. On the one side you have the supporters of nuclear energy busy claiming that it is by far the cheapest way of producing electricity, while on the other side you have the people who oppose it and who, of course, say the exact opposite. Who, then, is to be believed?

The first thing we have to remember is that the generation of nuclear electricity, just like hydro, tidal, geothermal and wind electricity, needs considerable capital expenditure before any power is produced, but it is then able to keep costs down by using a cheap source of fuel. The opposite is true for open-cycle gas-fired power stations, which need low up-front investment, but then have to face high fuel costs as their thermal efficiency is fairly low, at about 30-45%. Oil-fired, clean coal and combined cycle gas-fired power plants (with a thermal efficiency of 55-60%) come somewhere in between the two.

Things also get more complicated when one has to consider that there are different technologies in use both in fuel and coal-fired power stations, according to their thermal efficiency and the amount of pollutants and greenhouse-effect gases (including CO_2) they give off. To work out the cost price of a kilowatt hour of electricity for each type of power plant, we have to add up all the different parts that make up the cost price.

A simplified list of these costs comprises:

- capital costs
- fuel costs
- other operating costs
- insurance costs
- waste costs
- other external costs (including CO_2 emissions)
- dismantling costs

Before the electricity actually reaches the end-users, we have to add on additional costs for every type of power plant, such as marketing, transport and distribution costs, as well as tax. Even when reduced to its simplest form, the list is still long. We are now going to go through each of these costs so we can understand the difficulties involved in comparing the cost prices of electricity generated by the various types of power plant. Generally speaking, except for nuclear plants, we only need to take into consideration the first three elements, without, however, neglecting the other four.

Capital Costs

If we want to make a meaningful comparison of the capital costs of coal and oil-fired plants, hydro and wind plants, nuclear plants, etc., then we have to know their specifications-including their rating, the technology used, their efficiency, their date of construction and the country, as well as the annual number of hours they are designed to be used for or are actually used.

Once we have got these specifications out the way, we run into another major problem: how can we compare capital cost prices per kilowatt hour for plants which are not all operated for the same number of hours per year, or which have different uses or different useful lives? To keep things simple, we will limit the discussion to baseload power plants.

In power plants which are voluntarily used only part of the time, there are periods of peak need where the cost prices are generally higher, but where the electricity generated also has a higher economic value or selling price, which again tends to cloud the issue.

As nuclear power plants worldwide are generally only operated at baseload, this simplification can be justified.

However, this does not bring us any closer to addressing the second issue, the theoretical and actual service-life.

How can we compare capital cost prices per kilowatt hour for plants which are not all operated for the same number of hours per year, or which have different uses or different service lives?

The Eurostar High Speed Train at Waterloo station in London: For the French part of the line, 90% of the electricity used is non-CO_2 emitting

Issues concerning power plant service-lives and the discount rate to be applied play a major role in estimating the cost price of the electricity

When the decision is taken to build a power plant, it is designed to operate for a certain number of years, currently between 30 and 60 for nuclear plants. However, the theoretical plant service-life can be affected by a number of technico-economic factors that intervene during the plant's operating lifetime, such as corrosion, metallurgical changes, rise in maintenance costs and safety upgrading and it could turn out to be shorter or longer than planned.

The first reactors often failed to reach their forecast service-lives for the simple reason that the introduction of more efficient technologies made it more economic to shut them down (economic obsolescence). This was especially true for graphite-gas reactors, which made up the first generation of electricity-generating reactors and which became obsolete once the first PWRs came onto the market. In the United Kingdom, on the other hand, the British Magnox reactors saw their useful lives extended, as the government decided not to replace them with second-generation reactors, with the single exception of the Sizewell B PWR.

Most of the time, the service-life of the earlier reactors has been extended. In the United States, for example, in 2006, 40 nuclear plants which were licensed for operation for 40 years, had their licenses extended to 60 years and these are set to be followed by a good number of other plants. The issue of plant service-life is highly important when it comes to assessing the cost price of the electricity: the longer the planned service-life, the lower the capital cost per kilowatt hour produced.

The initial license term for a power plant (generally 40 years in the USA and elsewhere) reflects the amortization period set out by the operators for the capital investment. When the license term is extended and the capital investment has already been amortized, the cost price per kilowatt hour generated is going to be vastly reduced for the whole of the additional operating period. Costs will work out two or three times lower than for the unamortized, or full costs, which means that these power plants will have a substantial competitive edge over the other kinds of thermal plants. In the USA, it is common for power plants to change owners during their service-life and the cost price of the kilowatt hour will depend here on the purchase price of the plant. Most of the plants in this category in the United States enjoy very high profitability, but for artificial reasons.

Once the basic specifications have been set out (annual number of operating hours, plant service-life), the financial costs - the interest rate and the discount rate - need to be factored in. We know that a euro or a dollar is worth a lot more than today it will be in 10 years or so, but we are often unable to reach agreement over how much more it will be worth. What we do know is that the higher the capital investment and the longer the plant takes to build, the larger the impact the discount rate will have on the capital cost price per kilowatt hour generated. The effect of the discount rate used is shown below for three types of power plant.

It is clear that high discount rates benefit plants with technologies that require low capital investment and which can be built relatively quickly (gas, oil and coal-fired plants), while those with high capital

Effect of Discount Rates	Discount Rate of 5% or kWh Cost Price Base	Discount Rate of 10%
Nuclear Power Plant	100	150
Coal-fired Power Plant	100	130
Gas-fired Power Plant	100	110

Source : IEA and OECD-NEA 2005

When the license term is extended and the capital investment has already been amortized, the cost price per kilowatt hour generated is going to be vastly reduced for the whole of the additional operating period

Hoover dam on the Colorado river (Nevada).

Although the energy costs nothing, hydroelectric power has high up-front costs and can have a highly negative impact on the environment.

costs and long construction times, such as hydroelectric, clean coal and nuclear plants, will find themselves at a disadvantage. The cost of money, as expressed by the interest or discount rate, is therefore central to cost price calculations for comparing the cost of electricity generated by the different types of plant.

Unfortunately, there is no simple rule of thumb to follow for calculating the relative costs of the different technologies available. States, for example, when they decide to build railways, roads, canals and ports, etc. will be able to get low interest rates of about 5% in constant monetary units, to which is added the prevailing rate of inflation to obtain the interest rates in current monetary units. Private companies, on the other hand, are not only under pressure from their shareholders but also cannot get their hands on capital (equity or borrowed) as easily as States, which means they will have to start off with much higher interest rates, let's say at around 10% in constant monetary units. As can be seen, there is no single catch-all solution to the issue of the money rate to be selected when computing cost prices.

In France, the General Directorate for Energy and Raw Materials (DGEMP) of the French Ministry of the Economy, Finance and Industry, carries out periodical reviews of the cost prices of electricity generated by the different types of power plant. The last such review, carried out in 2005, gave a discount rate of 8% in constant euros (with a variant at 5%). This is the rate we will be taking, although private companies would be perfectly justified in taking higher rates, such as 10% or even 12%.

The table below summarizes the unit capital costs, based on the DGEMP figures and the French *Commissariat au Plan*, updated to reflect 2006 conditions. The costs represent the capital costs per thousand kilowatt-hours, or Megawatt-hours (MWh) generated by baseload plants in France.

The table shows that the technology requiring the highest capital investment per unit of electricity generated is wind

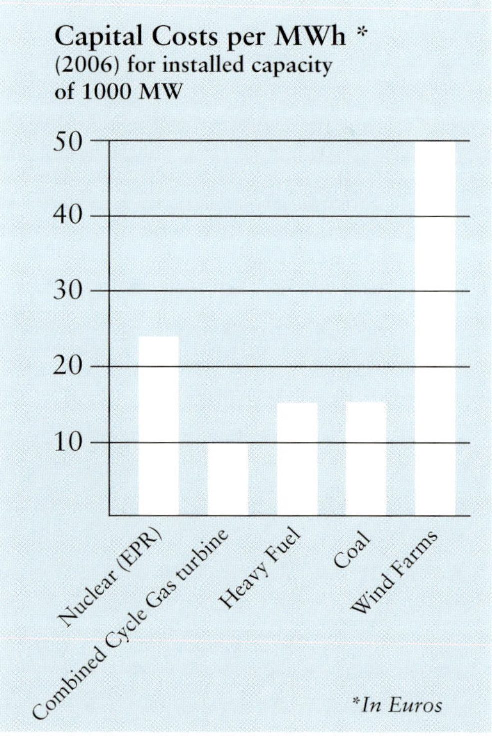

power. This is because on average during the year, a wind farm in Europe has a capacity factor (the amount of energy a facility generates in one year divided by the total amount it could generate if it ran at full capacity) of 20%. Projections for a wind farm in France in the future are based on a capacity factor of 30%, or 2,500 hours of full production per year, which is much more optimistic.

Nuclear power plants have the second-highest up-front costs, but they operate with a capacity factor of between 80 and 90%, a figure that is the same in France, Germany, Russia, Sweden, Switzerland, China and the United States. Of the thermal power stations, nuclear is the most capital-intensive and the capital costs work out three times higher than for combined-cycle gas turbines and some 30% higher than clean coal plants.

It would be difficult to make accurate comparisons on an international scale as construction costs and lead times vary considerably from one country to another, as can be seen in the table at the bottom of the page. The figures allow for the sharp rise in the prices of steel and other raw materials between 2003 and 2006.

Fuel Costs

Fuel costs of the different types of power plant are another stumbling block when it comes to calculating the comparative cost price of electricity. When the plants rely on renewable energies such as wind or hydroelectric power, the fuel costs are, by definition, zero. However, when we consider other types of renewable energies such as biomasses (wood, agricultural waste), which have to be grown and subsequently collected, these costs are much higher. The most significant comparisons are those to be made between fossil-fuel power plants, running on gas, oil or coal, and nuclear plants. The main problem in comparing the fuel costs is the fact that fossil fuel prices vary considerably with time, even if they are all more or less linked to the price of crude.

When fuel prices are computed, we need to know what these calculations are based on: what is the price of crude (which will give us the price of oil), the price of gas and the price of a ton of coal for a power plant?

New York Stock Exchange.

Investment Costs in $G for 1,000 MW (2006 Conditions)

1 Euro = 1.25 USD	Nuclear (EPR Type)	Combined-Cycle Gas Turbine	Heavy Fuel Oil *	Coal	Wind Farm
USA	$2,5 G	$0.7 G	$1.2 G	$1.2 G	$1.5 G
EU (France)	$2.5 G	$0.7 G	$1.5 G	$1.5 G	$1.5 G
Japan	$3.0 G	$1.2 G	$2.0 G	$2.0 G	n.d.

(*) Plants now being phased out

Wind energy costs nothing, but for each kW capacity, a windmill needs twice as much concrete and 3 times more steel, while for each kWh produced, it needs 8 times more concrete and 12 times more steel than for a nuclear power plant.

The table below gives two cases which provide interesting comparisons. The first case is for the year 2005, with the price of crude at $50/bbl, while the second is a projection for 2020 with the price of crude put arbitrarily at $100/bbl. We have done the same thing with uranium, taking the uranium resources and reserves as described in chapter 4, which gives us $65/kg for 2005 and $130/kg for 2020.

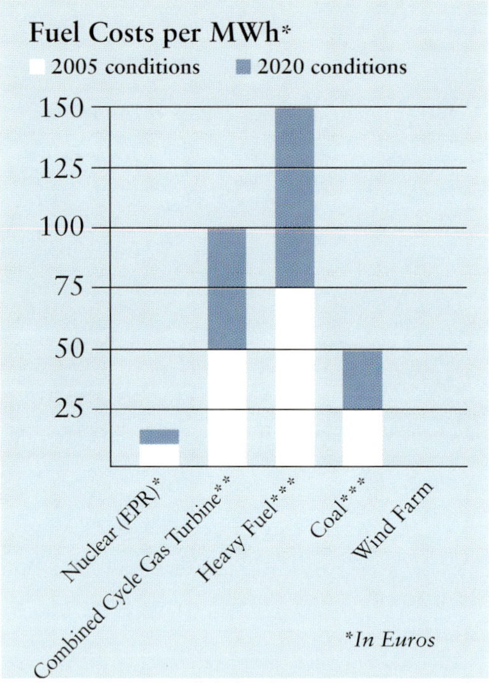

(*) For nuclear fuel, the raw material amounts to under 50% of its cost, the other half coming from its conversion, enrichment and manufacture (pellets of uranium oxide and fuel cladding). This explains why if the cost of uranium were to double by 2020, the cost of nuclear fuel would only rise by some 50%. The net cost of reprocessing can currently be considered as negligible and even negative.
(**) Efficiency 60%
(***)Efficiency 45%

Other Operating Costs

Like in any production facility, apart from the raw material needs (which is the fuel in nuclear or conventional thermal power plants), the operating costs cover all the other expenses incurred in running the facility:

- labor costs
- maintenance and spare parts
- consumables

Insurance costs generally come within operating costs, but they will be described separately here as the nuclear industry is often criticized for not factoring in its insurance costs. Similarly, decommissioning, dismantling and cleanup costs will also be given separately.

One common feature of electricity power plants of all types is that operating costs are low compared to the two major cost items, the capital costs and the fuel costs.

For this reason, we will not go into the details of the operating costs, but will give a highly simplified picture, using the orders of magnitude given below, in euros per MWh:

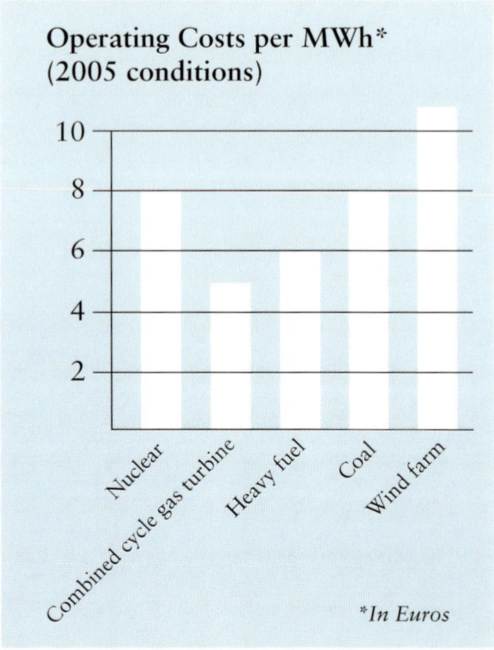

A first approach to Electricity Cost Prices according to Type of Power Plant

We have already put a figure to the three main items that make up the cost price of electricity produced by the various types of power plant and these can be set out in the form of two tables:

Despite all the unknowns that surround this sort of projection, it is worth making a second table to see if nuclear power will still be competitive around the year 2020.

The biggest change in this table compared to the first one will be the sharp rise in the cost of fossil fuels and CO_2 emissions.

Cost Price of Electricity produced by the various Types of Power Plant

2005 Conditions Price of Oil $50 /Barrel	Nuclear	Gas	Heavy fuel	Coal	Wind Farm
Capital Costs	20	10	15	15	50
Fuel Costs	10	50	75	25	0
Operating Costs	8	5	6	8	12
TOTAL	**38**	**65**	**96**	**48**	**62**
CO_2 Costs at $20/t	**0**	**7**	**10**	**15**	**0**
TOTAL COST WITH CO_2	**38**	**72**	**106**	**63**	**62**

The first table clearly shows the competitive edge of electricity generated by baseload nuclear power plants (large plants of 1,000 MW minimum). This certainly does not mean that there is no room for other types of power plant, as we can have plants that can be used as semi-baseload or peakload generators or others that need to be subsidized over an initial period so as to create a market that will help them reduce costs through economies of scale and technological progress. Examples of the latter case include renewable energies, except for large hydroelectric projects.

If the projection in table 2 turns out to be anything near reality, simple economics should result in a rapid development of nuclear power and energy conservation in all forms. The situation should also see development of wind power although this will probably not be on a very large scale, as a nuclear power plant would be too slow to respond efficiently to compensate for the variability of wind power. Nuclear plants would also be the most expensive backup one could find for producing power on a peaking basis. Wind farms therefore need to be built with a gas-turbine backup system of about a third of their size. This

The table below shows that the rise in fossil energy prices, resulting mainly from worldwide gas and oil production peaks, and the rise in the cost of CO_2 emissions is going to have a profound effect on electricity production in the world.

Cost Price* of Electricity produced by the Various Types of Power Plant

(2020 Conditions) **Price of Oil $100 /Barrel**	Nuclear	Gas	Heavy fuel	Coal	Wind Farm
Capital Costs	20	10	15	15	40
Fuel Costs	15	100	150	50	0
Operating Costs	8	5	5	10	10
TOTAL	**43**	**115**	**170**	**75**	**5**
CO_2 Costs at $100/t	**0**	**35**	**50**	**75**	**0**
TOTAL COST WITH CO_2	**43**	**150**	**220**	**150**	**50**

**In Euros per MWh*

explains why their development will be somewhat restricted and it also means that their impact on CO_2 emissions will be lower, as over the year, the gas-turbine backup, a CO_2 emitter, will be operating about 75-80% of the time, if the wind farms have a conventional capacity factor.

Wind-power in Denmark accounts for 20% of its electricity production, the highest in Europe, but it is also one of the highest CO_2 emitters per kWh produced (the rest comes from coal)

Insurance Costs

No analysis of the cost prices of electricity generated by the various types of nuclear plant would be complete if we did not factor in the cost of insurance.

Critics of nuclear power often complain that the cost-price of nuclear-generated electricity fails to include the cost of insurance against accidents. There are, however, two sides to this argument.

First of all, there are the countries like the United States where the owners of nuclear power plants look to the market for insurance to cover against nuclear accidents, with claims capped at, for example, one or two billion dollars. The insurance companies themselves reckon that as the actual risks are so low, the nuclear sector is one of the most profitable ones. If the damage goes over this initial cap, the bill is paid out of the public purse. In this kind of situation, the scale of the accident means that it is considered as a natural catastrophe. It should not be forgotten that the same conditions apply to hydroelectric dams and that there have so far been a lot more deaths caused worldwide by dams breaking than by nuclear power-plant accidents.

A spent fuel assembly being loaded underwater into a TN 12 transport package.

In other countries such as France, producers of nuclear electricity prefer to act as their own insurers, given the very low probability of a major accident affecting large populations.

Since EDF was partly privatised in 2005, it has looked to the market to insure part of its nuclear risks.

Waste Disposal Costs

Waste-disposal costs cover two aspects: the cost of reprocessing in those countries which have opted for this solution (the closed-fuel-cycle chosen by Germany, France, the United Kingdom, Japan, etc.) and the cost of interim storage and final disposal of the waste.

The actual net cost of reprocessing is a complex issue as we have to factor in the costs incurred, the value of the recovered finished products (reprocessed uranium, plutonium) and the reduction in waste volumes and subsequently the costs of disposal after reprocessing.

Such a calculation does not come within the scope of this book and we will just make two brief points.

- firstly, there is no "right" answer, as the value given to plutonium depends entirely on the long-term development of nuclear power. It will have no value at all or even a negative value for countries who have decided to abandon nuclear power, while for countries who have opted for the superbreeder solution, plutonium is an essential component for sustainable development and will therefore have a high or very high value,
- secondly, the value of the reprocessed uranium needs to be determined, and this varies in the same way as mined uranium, which makes it very difficult to pin down, especially given the spectacular rise in the price of the metal since 2003.

To summarize, and just to give an idea of scale, for a closed-fuel-cycle, reprocessing added about 10% onto the cost of nuclear fuel under conditions in the year 2000, but had a negative cost under 2006 conditions.

For those countries which have opted for the fast-breeder solution, the cost of reprocessing is already strongly negative. Reprocessing is necessary for the long term and is a highly profitable activity. Putting a figure to this negative cost and therefore estimating the earnings from reprocessing will be largely dependant on the prices depleted uranium and plutonium are expected to fetch, as for the moment there is no market for these products.

The United States, the country that consistently supported the open fuel cycle and where the fuel is not reprocessed, has

Vitrification of most radioactive waste prevents it being dispersed and reduces its volume, making it easier to store.

started to backtrack over the last few years and is now actively considering switching over to reprocessing. Japan opened its first commercial reprocessing plant (Rokkasho Mura) in 2006.

Whether the fuel is reprocessed or not, there still remains the question of the cost of interim storage and ultimate disposal of the waste. the technical points of which are discussed in chapter 11. Economically speaking, the question is not a real issue. If we compare these costs to the cost price of a kWh of nuclear electricity, they turn out to be very low: 1% of the cost of the fuel (the optimistic figure), making 0.3% of the cost price of the kWh. Even if we were to go for the most pessimistic scenario with a tenfold increase in the costs, the result would come out at 10% of the price of the fuel, or 3% of the cost price of the nuclear kWh.

The storage and disposal of nuclear waste, although of paramount importance from a safety point of view, does not have such a big economic impact.

The other External Costs

The only significant external costs for nuclear power are the human costs of a nuclear accident and also the costs involved in conditioning and storing the waste, funds for which have already been set aside. When one looks at the external costs of fossil-fuel power plants (gas, oil, coal), one thing that stands out is that no costs relating to atmospheric pollution (SO_x, NO_x, dust) and in particular CO_2 emissions have as yet been factored in.

Climate change is becoming an increasingly important factor in our future energy needs (see chapter 14) and is now a central part of issues relating to sustainable development.

Economists consider that all the methods that lead to a reduction in CO_2 emissions (taxes, tax incentives for improving energy efficiency, emission caps, emission-rights trading) boil down to a simple question of what is the cost to be allocated to a ton of carbon in the atmosphere or what is the cost of emitting a ton of CO_2 (a ton of carbon is the equivalent of 3.66 tons of CO_2).

What is the cost to be allocated to a ton of carbon in the atmosphere or what is the cost of emitting a ton of CO_2: $20 per metric ton in 2005, but $50 or $500 per ton in 2020

The cost of carbon dioxide emissions does not exist as such, but is part of CO_2 reduction legislation (the Kyoto protocol, national and international law and also State and local law in the USA). All the comparative studies that include the external costs of CO_2 emissions must at present be considered as mostly arbitrary and liable to undergo considerable change in the long term. With climate change becoming

Change in CO_2 emissions over the last 1,000 years.

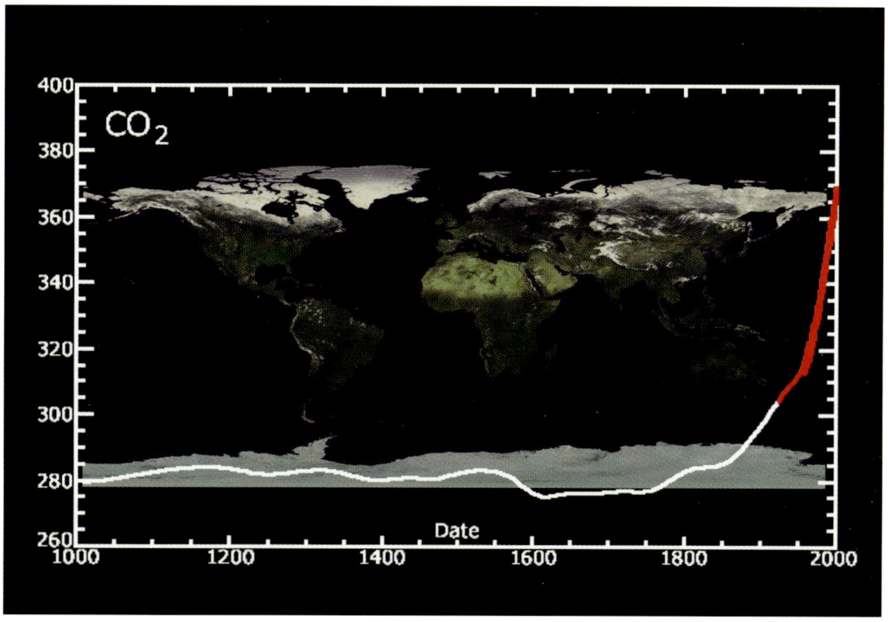

an issue of increasing concern, the restrictions will become tougher and tougher and the costs of CO_2 emissions will rise proportionally. If these restrictions are applied worldwide, the cost of the emissions will become increasingly evenly-balanced.

In the two tables on page 109, we compared the impact of the rise in the cost of CO_2 emissions between 2006, when it stood at $20/t for Europe, and 2015-2020, where we projected what we believe to be a realistic cost of $100/t for Europe, the USA and Japan, and possibly also for the emerging countries. The table shows that the increase in CO_2 costs has a significant impact on the total cost price of electricity generation in three out of the five technologies under consideration.

Nuclear power, which was already the cheapest form of energy without taking into account the cost of CO_2, now has a clear competitive advantage over the other forms (except wind-power) and this gap will continue to grow as CO_2 emissions become more and more expensive.

Although the costs involved in dismantling look extremely high in absolute terms, they actually only account for a very small proportion of the cost price of the nuclear kWh

It has often been argued that the cost price of nuclear energy should also allow for external costs relating to plutonium production. However, as we saw above, for countries which have opted to follow the superbreeder path towards sustainable nuclear energy, far from having a negative value, plutonium actually has a highly positive one, which means the argument would be economically meaningless. There are nonetheless substantial costs that are difficult to assess properly. These are the decommissioning and dismantling costs for nuclear plants at the end of their service-lives.

Dismantling end-of-life installations

The costs involved in decommissioning and dismantling nuclear power plants of any kind are high, as the plant components are radioactive to varying degrees.

Dismantling, therefore has to be performed with extreme caution, which means it will take a long time and be costly. However, the costs are certainly not unknown or disproportionately huge. The nuclear industry has accumulated considerable experience in dismantling nuclear power plants. In most countries now, these costs are included in the cost price of nuclear electricity. In France, the French electricity utility, EDF, is legally obliged to include provisions for dismantling and to set aside funds for it.

The question that comes up is whether these funds are actually sufficient. They currently amount to about 15% of the construction cost, which means that even if they have been underestimated by 100%, they would still not have a major effect on the cost price of electricity and their impact would go up from the current figure of 2% to about 3 or 4%.

It can be seen, therefore, that although the costs involved in dismantling look extremely high in absolute terms, they actually only account for a very small proportion of the cost price of the nuclear kWh. What is important is that companies and States make sure that provisions have been set aside to allow for decommissioning and dismantling costs and that this money is kept solely for this purpose.

As the cost of decommissioning and dismantling also depends on the level of residual radioactivity in the plant components, it is worth waiting as long as possible before beginning the process: the "colder" the plant is, the less radioactive it will be and the cheaper it will be to dismantle. This is an interesting point for discussion between the generations, which also touches on sustainable development: how much time, from the economic, technical and ethical points of view, should elapse before dismantling is begun? What period would be the most politically acceptable for public opinion?

Conclusion

It would be difficult to wrap up this analysis of the cost prices of the various types of electricity generation without stating that for the next fifty years at least, nuclear electricity will clearly be the cheapest and safest (from the geopolitical angle – see the chapter on storage costs) method for the large-scale baseload production of electricity. Renewable energies, especially wind-power, are the most competitive sources of energy for meeting semi-baseload or peak-load demand, although it is still difficult to estimate the overcosts involved in a discontinuous source of energy such as wind-power.

In the long term, solar energy might come into prominence, but the technology needs to move a few steps forward before it can become a meaningful player. The problem of the higher costs of intermittent electricity could be solved if the industry manages to come up with a solution for the large-scale storage of the electricity. Hydraulic pumped storage, currently the only operational method, is only feasible in a few high-mountain sites.

The most expensive energy sources and those which present the biggest danger for the environment, are those that use fossil fuels, especially coal. Environmental concerns mean that the capture and storage of CO_2 is going to become a major issue and if it turns out that the large-scale of storage of CO_2 is economically unviable, then the momentum that will have gathered for the development of nuclear energy is going to be very hard to stop.

To get back to the present day, it is worth noting that the ex-tax price of electricity in France (€53.3/MWh for industry in 2005 was about half of the price in Italy (€101.8/MWh), where there is no nuclear power). The figures owe nothing to chance, but reflect economic reality. Nuclear power is not just set to be the cheapest form of electricity in the future, it already is now and the proportion of electricity generated by nuclear plants is going to be a key factor in both economic competitiveness and the standard of living of both the world's leading and emerging countries.

To finish off this chapter outlining the economics of nuclear energy production, we can recap the paradoxes we have tried to explain and which are the reason why nuclear power seems to be so full of contradictions. These are some of the points that should be remembered when comparing nuclear plants and other types of power plant:

- nuclear power plants are the most expensive to build per unit of installed capacity (cost per kWh) and take the longest to build,
- nuclear power plants operating at baseload generate electricity at the lowest cost (cost per kWh),
- nuclear power plants have much lower variable production costs than coal, oil or gas-fired thermal power stations,
- nuclear power plants have the lowest external costs.

The economic features of nuclear electricity generation are very much like those involved in the hydroelectric production of electricity, but completely different to those involved in thermal power plants using fossil fuels.

In the long term, solar energy might come into prominence, but the technology needs to move a few steps forward before it can become a meaningful player.

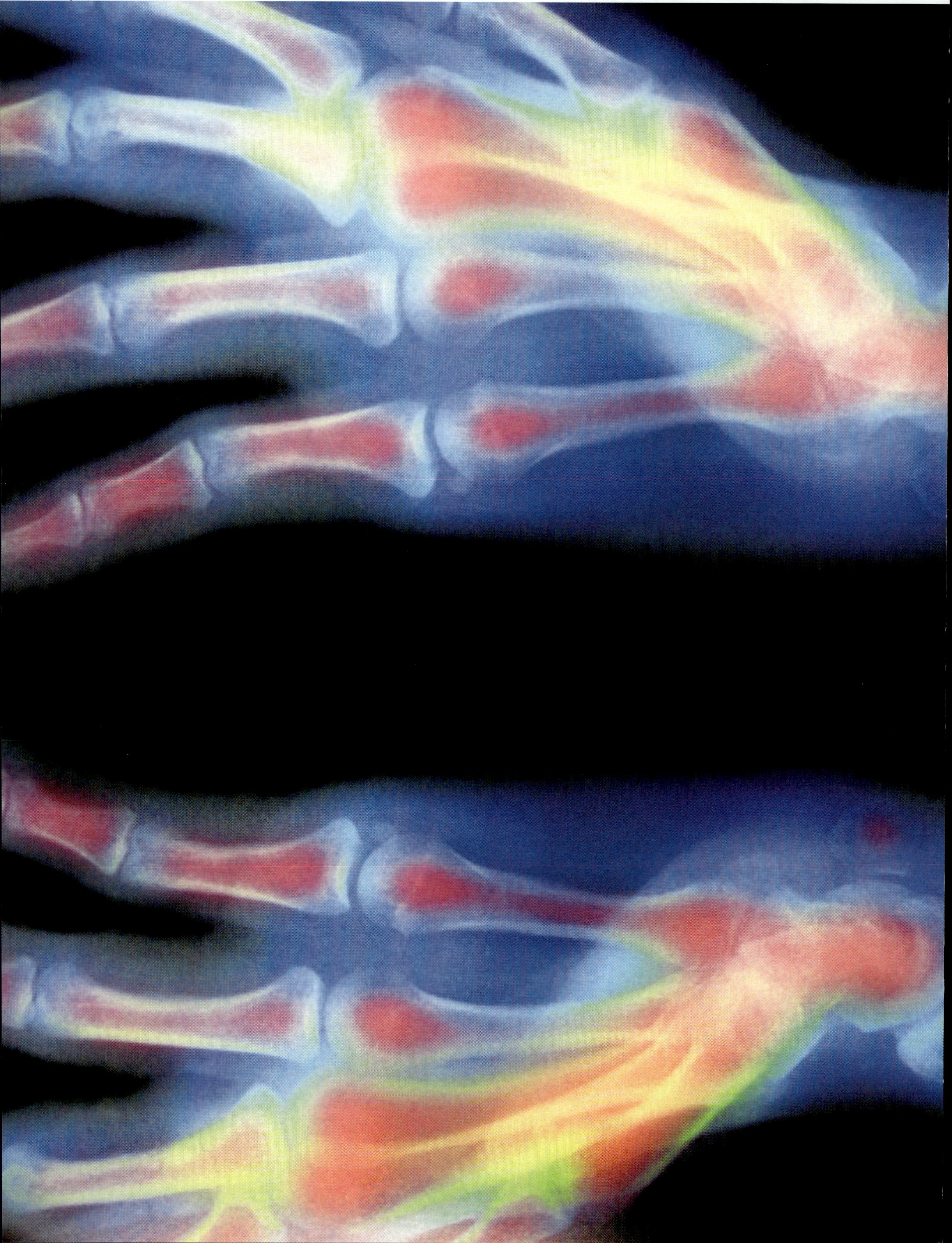

Other Applications of Nuclear Energy

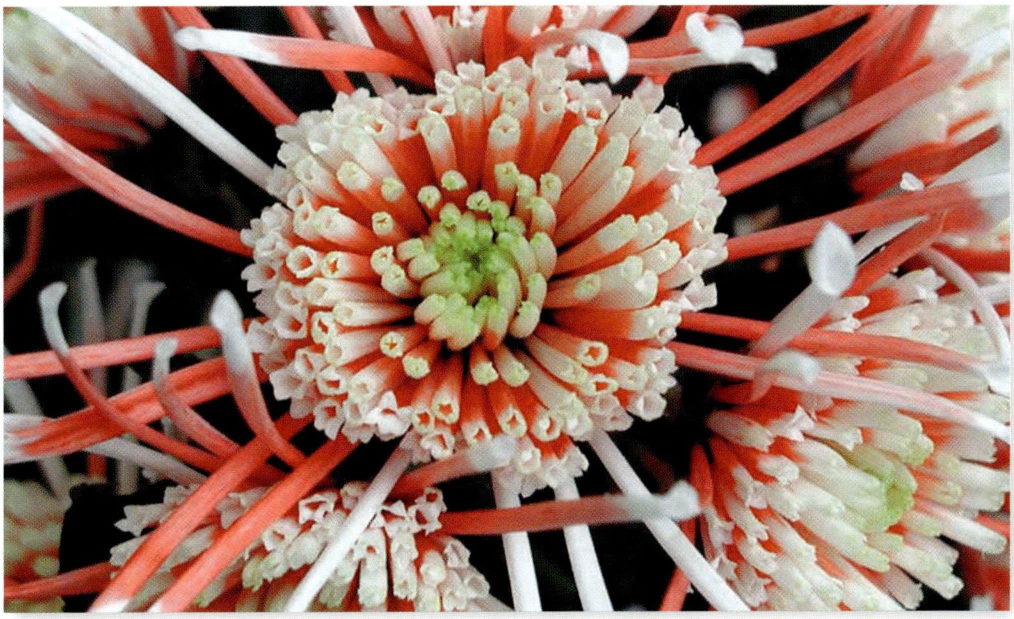

Health & Medicine	117
Agri-food Applications	120
Various Industrial Applications	123
Naval Propulsion	125
Space Applications	126
Environmental Protection	128
A limitless Field of Applications	129

Nuclear and radiation techniques are, however, used in all these fields and often play an essential role

When anyone mentions the word "nuclear", it immediately conjures up images of power plants or, perhaps, atomic weapons, but not many people would think of medicine, agriculture, food and manufacture industries, environment or space.

Nuclear and radiation techniques are, however, used in all these fields and often play an essential role.

Perhaps even more surprisingly, studies carried out in the USA and Japan have shown that the total revenue generated by civilian non-power applications of nuclear and radiation techniques was significantly larger than that produced by nuclear power.

Direct Economic Impacts in 1997 (billion US$)

Uses of radiation:	USA	Japan
Health, Medicine	49	12
Agri-food	14	1
Industry	56	39
Sub-total	119	52
Nuclear Electricity	*39*	*47*
TOTAL	158*	99
%PIB	1.9%	2.3%

* $ 420 billion, indirect impacts included

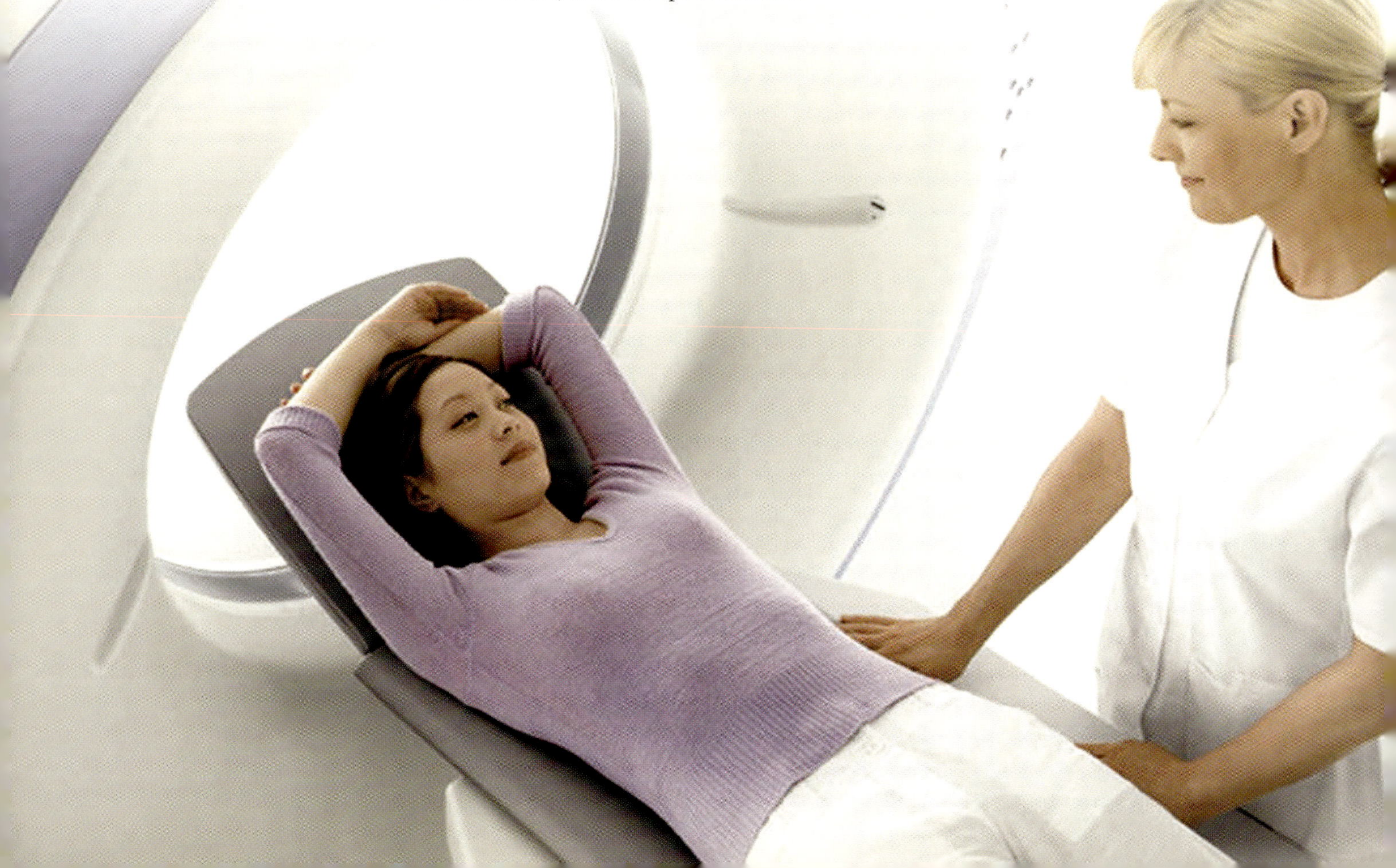

Health & Medicine

Nuclear techniques are now widely used in medicine, for instance :

- Genomic research
- Medical imaging and functional exploration
- Radiotherapy (gamma, neutrons, heavy ions)
- Curietherapy
- Targeted irradiation
- Sterilization of surgical tools
- Biocompatible materials
- Wiping out parasitic illnesses...

Nuclear technologies applied to medicine have helped cure hundreds of millions of patients. The first time you visit a biotechnology company you cannot help being surprised by the number of times you see the distinctive radioactive materials sign on the doors.

Without isotopic tracers, using stable or in most cases radioactive isotopes, there would be no modern biology, and the genome would never have been sequenced. Without radiotherapy, cancer would be an even greater killer than it is today.

Without isotopic tracers, using stable or in most cases radioactive isotopes, there would be no modern biology, and the genome would never have been sequenced

Medical Imaging

Medical imaging technology has expanded enormously since the first X-ray machines were built and now includes ultrasonic scanning (a spinoff from asdic sonar research) and nuclear magnetic resonance imaging.

Gamma densitometry and PET tomography are both derived directly from nuclear technology, or more specifically from radiation detection and measurement:

PET Tomography (Alzheimer)

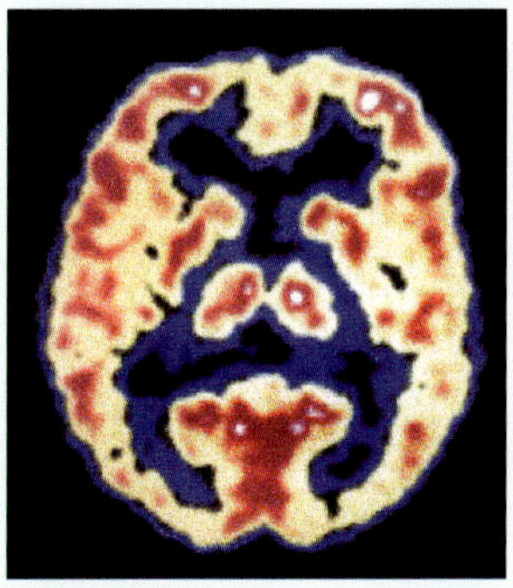

Normal brain

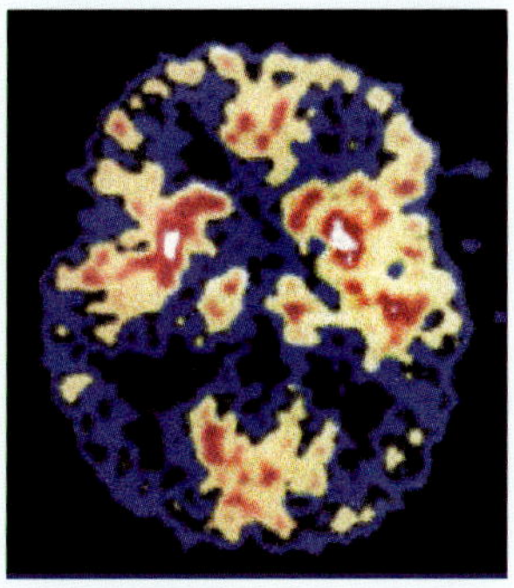

Brain affected by Alzheimer's disease

The principle of PET (Positron Electron Tomography) is fairly complex: the patient is injected with a substance (sugar or a drug) marked by a short-lived $\beta+$ emitter.
When a nucleus of this marker decays, it emits a positron which very soon runs into a "normal" electron.
The two particles annihilate each other and emit two diametrically opposed γ rays : the length travelled by each γ photon before reaching a ring detector can be measured by time-of-flight to find the emission point, and hence locate the radioactive marker.
This means we are able to locate the zones within the brain activated by specific brain activity, as these areas consume the sugar carried by the blood. The technique makes it possible to track the metabolites of a given drug after it has been absorbed.

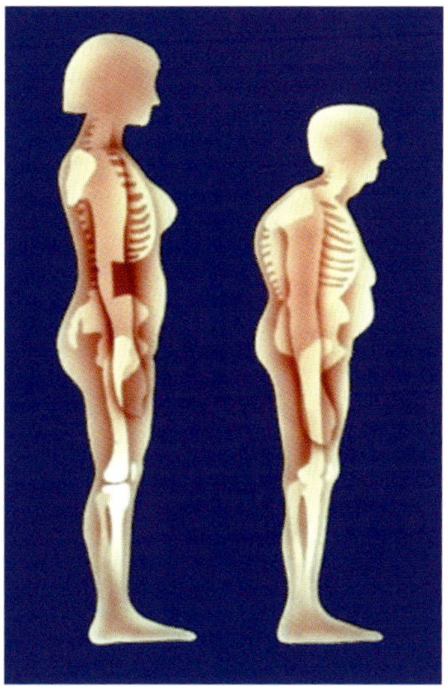

Gamma Densitometry (Osteoporosis)

Gamma densitometry involves using a gamma-camera in the diagnosis of osteoporosis by providing precise measurements of bone density through differences in gamma ray absorption. Osteoporosis is a loss of calcium in the skeleton which affects mostly post-menopausal women and makes patients prone to bone-fractures.

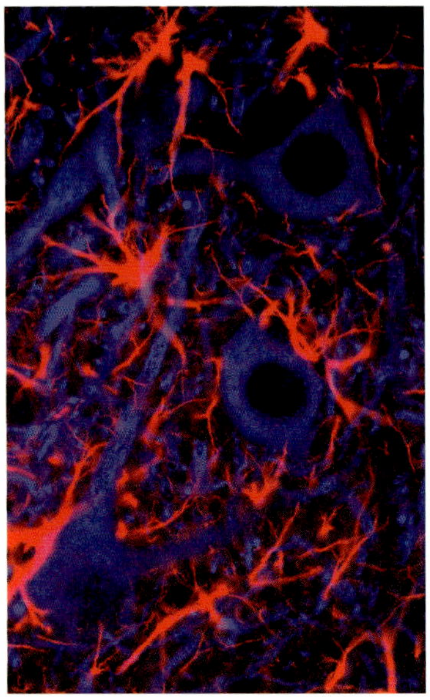

Astrocytes are star-shaped cells surrounding the neurons in the brain. They are thought to play a role in Alzheimer's disease.

Composite Image of a Lung Cancer Patient

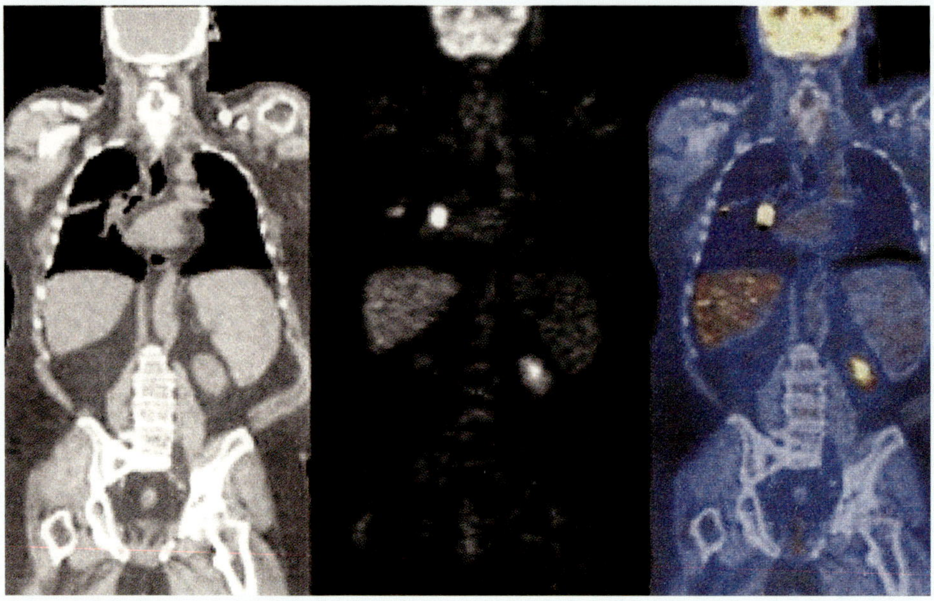

Imaging can be enhanced by merging information from several detectors, such as an X-scanner and PET or MRI and PET and combining them on a single picture.

The destruction of solid tumours by X rays has been replaced by γ-radiotherapy

Therapeutic Irradiation

The very first application of radioactivity was medical. Following the work carried out by Pierre and Marie Curie, radium sources were used to treat skin lesions, lupus and skin cancers as far back as 1901, although with mixed results. By 1920, however, curietherapy had become the most efficient way of treating cervical cancer. The technique is still used today, but radium has been replaced by artificial elements that are easier to use, such as iridium wires or iodine 125 encapsulated in titanium, which has proved to be efficient in the non-surgical treatment of prostate cancers (it is commonly called brachytherapy, as the irradiation is intense but the affected area very small).

Although X-rays had long been used to destroy tumours, a significant breakthrough was made in 1951 with γ-radiotherapy and the cobalt bomb, which bombarded malignant cells with gamma rays from several different directions. The technique was particularly effective against deeper-lying tumours, while its greater precision (the gamma scalpel) meant it caused much less damage to surrounding tissue than conventional X-ray treatment.

Radiotherapy can also involve the use of accelerator-produced radioactive isotopes or neutron beams fired from an experimental reactor. In the same family, BNCT, Boron Capture Neutron Therapy,

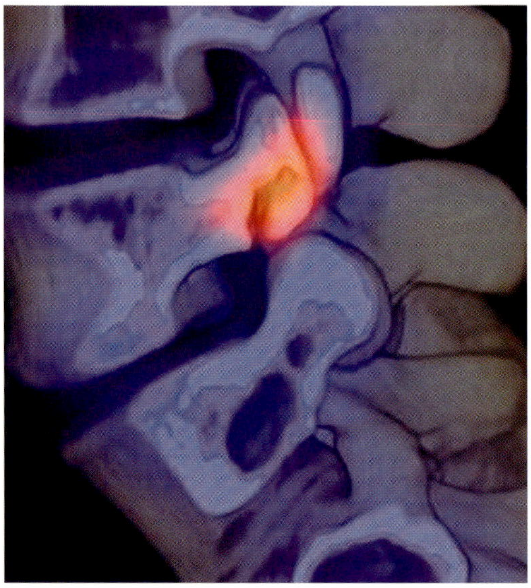

Treatment of a malignant Nasal Cavity Melanoma by heavy Ion Irradiation

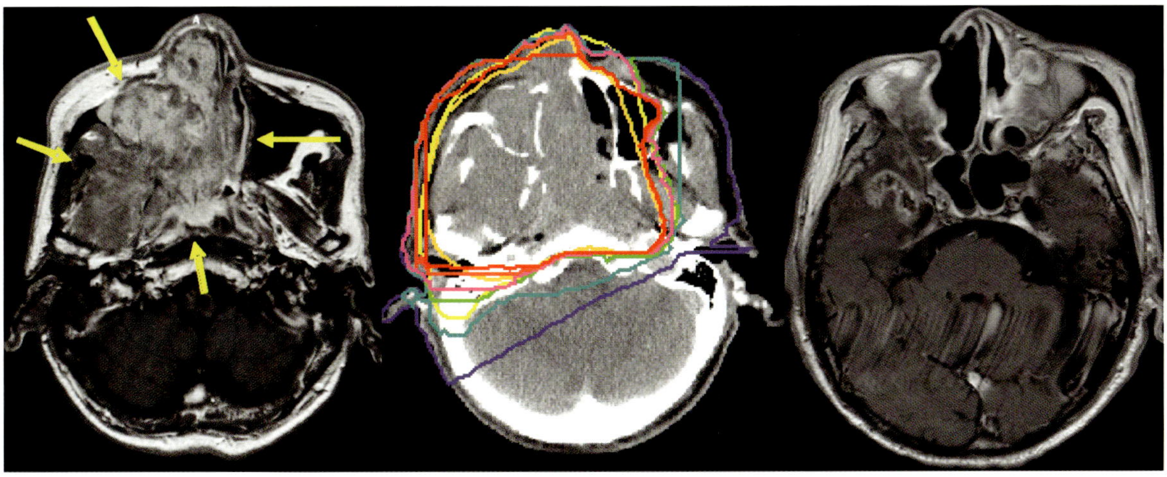

HIMAC facility in Chiba (Japan).

attaches a stable isotope of boron (10B) to a tumour-seeking compound. The tumour site is then irradiated by neutrons, which cause the boron to eject short-range α particles which will destroy neighbouring tumour cells, with minimal collateral damage

Another technique, which is still at the experimental stage, involves grafting a short-lived α-emitter (bismuth 213) on a monoclonal antibody able to recognize and specifically bond to malignant cancer cells

A discussion of the direct medical applications of nuclear techniques would not be complete without mentioning the sterilization of surgical instruments and implants as well as the biocompatible hydrogel dressing manufactured by radiation-induced cross-linking of water soluble polymers.

One final technique worth examining is SIT, which involves rearing large numbers of insect pests and then irradiating them before hatching to sterilize the males, which are then released into infested areas..

The technique was initially developed to eradicate insect pests that were destroying crops and attacking livestock (see next page). It is now successfully being used to combat human parasitic diseases One example of SIT in action is the eradication of the tsetse fly from Zanzibar under an IAEA-sponsored irradiation program.

Nuclear techniques such as MRI mammography help in the diagnosis of breast cancer. Through a combination of surgery, chemotherapy and radiotherapy, most cancer patients can now be cured.

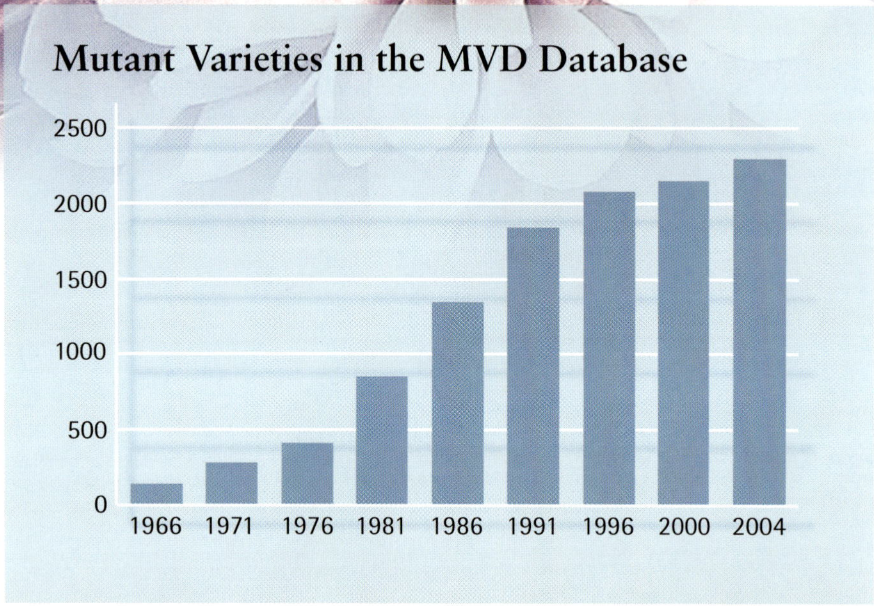

Agri-Food Applications

Ionizing radiation is widely used in the Food and Agriculture sector and already plays a key role in several fields, such the development of new plant varieties (cultivars), irrigation water management, food conservation through irradiation, pest and mould control.

Cultivars

Since 1966, over 2,300 plant mutant varieties have been officially recorded in the MVD (Mutant Varieties Data) database jointly managed by FAO and IAEA (see table, left) : 89% of those cultivars were developed through irradiation, including many varieties of staple food, vegetables, flowers and fruit.

The foodstuffs can be irradiated in the open with a gamma source, or inside, using an ion accelerator. The technique can also be applied to hydroponics (growing crops on a neutral substrate irrigated with a liquid nutrient).

Radiation-induced mutation can have spectacular results, such as higher crop yields for cotton (Pakistan) and rice (China, Egypt), improved storm-resistance in short stalked wheat (Italy), the last minute rescue of American mint (the basis of chewing gum) threatened by a fungus. About 27% of Chinese crops and 12% of Indian crops come from irradiated cultivars.

It is also possible to save on irrigation water by using neutron probes to measure in situ the soil moisture level at the plant roots.

New chrysanthemum colors.

Experimental field growing an irradiation-produced cultivar.

Food Ionization

Food irradiation or ionization is the exposure of food to controlled doses of ionizing radiation in order to:

- Kill pathogen bacteria, parasites and pests ;
- Keep products fresh by delaying oxidation, sprouting and rotting;
- Reduce waste and facilitate transportation, especially international transportation.

It is sometimes also called "cold pasteurization".

Food Ionization already dates back a long time. Authorized now in some 40 countries and tightly controlled, it actually started off in the 1950s, as part of the Atoms for Peace initiative.

The process was first used in Germany in 1957 to kill insects and mould in spices. In 1964, ionization was widely adopted to delay potato sprouting. In1970, NASA selected the process as the best way to sterilize astronaut's food without altering its taste. It was only in the 80s that authorization was obtained to irradiate meat - notably pork to prevent trichinosis (FDA 1985) - followed by fruit, vegetables and poultry (1992).

Egg irradiation was approved in 2000, with fish next on the list. The table below summarizes the irradiation doses.

Ionization makes hardly any alterations to foodstuffs, but it is not suitable for a number of fresh products, and it is sometimes possible to notice changes in taste. Nutritional value, however, is unchanged and the alterations as such are smaller than those caused by pasteurization.

It was only in the 80s that authorization was obtained to irradiate meat

Hydroponic culture in test-tube conditions.

2000 supermarkets sell irradiated beef in the USA.

Food Irradiation Specifications

Low Doses < 1 kGy	• Pest control in seeds and fruit • Delaying tuber sprouting • Delaying fruit and vegetable ripening • Eradication of parasites from animal products
Medium Doses 1 to 10 kGy	• Elimination of Salmonella, E. coli and listeria in meat and poultry • Delaying the onset of fungus (strawberries and other fruits)
High Doses > 10 kGy	• Destruction of insect and microorganisms in spices • Complete sterilization of food (e.g. for sterile bubbles)

The fight against Insect Pests

Insect pests can wreak havoc in crops and livestock and science has developed several techniques to combat them. The most efficient method, the sterile insect technique, SIT, dates back to the 50s and involves rearing a vast number of the male pests, irradiating them with γ rays (enough to sterilize them but not enough to remove their sexual drive), and then releasing them into the infested areas. This was the method used in America in 1966 to eradicate the screwworm.

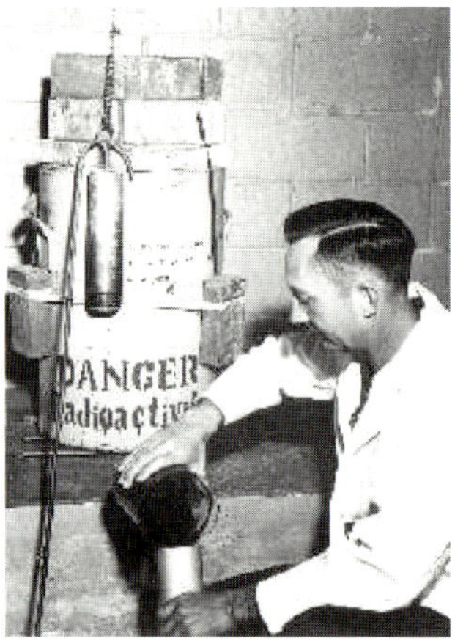

Screwworm eradication.

The Governor and the Fly, a Californian Story

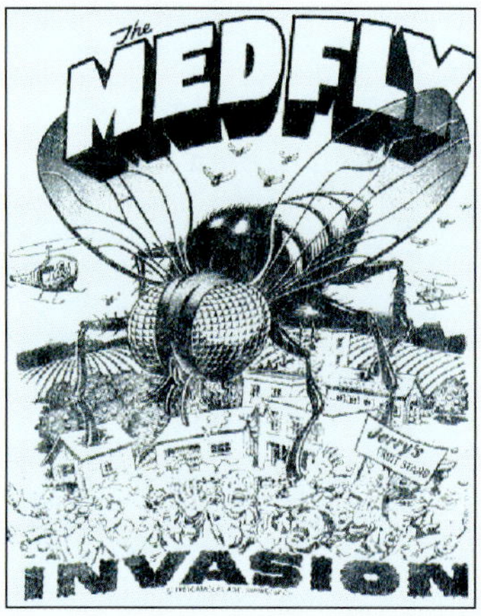

In the early 80s, Democrat Jerry Brown was Governor of California, the wealthiest State of the USA and which still had a strong hippy fringe. Governor Brown was strongly influenced by the ecology movement and decided to ban all pesticides from his State. The following season, all the orange plantations - of which there were almost as many in California as in Florida - were infested by the "Med Fly", which came over hidden in pineapple shipments from Hawaii. The med-fly plague was so acute that neighbouring States set up road-blocks quarantining California, adding insult to (economic) injury. To put an end to the havoc, California had to be given a thorough spraying with Malathion, the most virulent chemical pesticide available... For this and other unorthodox ideas, Jerry Brown gained the nickname Governor Moonbeam and lost the following election. In the end, the Med Fly was eradicated through the sterile insect technique.

Industrial Applications

There are so many industrial applications of radiation and radioisotopes that it is difficult to draw up a comprehensive list. Here are just a few:

- Polymerization and cross-linking of plastic and elastic materials
- Welding inspections, on line or in situ
- Measuring the thickness of steel and aluminium sheets
- Monitoring the bulk feed of processes using belt conveyers
- Fire detection
- Fight against terrorism (detection of explosives)
- Checking food package fill levels
- Industrial imaging (neutrons and gamma)
- Stratigraphic analysis for oil exploration

Most of the checking and monitoring applications (thickness, level, density) are based on measuring radiation absorption.

An Example: the Use of Nuclear Techniques in the Oil Industry

The oil industry is the sector with the heaviest investments and it makes wide use of nuclear technologies.

GEOLOGY
Methods for absolute dating of fossils and minerals and their inclusions.

GEOPHYSICS
Methods for characterizing rocks and the liquids they contain, during drilling operations. These measurements (called diagraphies or "logs" in oil jargon) use a variety of nuclear-based techniques:

- Neutron logs assess rock porosity by measuring the amount of hydrogen (water or hydrocarbons) in geological formations ,
- NMR logs (Nuclear magnetic resonance) or microscan tools,
- Gamma ray logs.

RESERVOIR STUDIES
Methods for assessing the rock fracture patterns through the use of radioactive tracers.

CONSTRUCTION
On-line check of welds during pipeline installation (oil and gas) and other non destructive inspections.

OPERATIONS
Level and flow detectors using nuclear techniques

MAINTENANCE
Radiography and other nuclear techniques for measuring corrosion and ageing of metal alloys.

IN THE FUTURE
Oil and nuclear industries are bound to become more and more complementary. They currently frequently make use of the same technologies, but this complementarity is set to grow as between them as the two sectors have to answer the world's energy needs (see chapter 14).

New Materials developed through Irradiation

Advanced Materials : Ceramic Composites

Silicon Carbide Fiber

Si-containing Polymer Fiber

EB irradiation up to 10MGy

Cured Polymer Fiber

Pyrolysis in inert gas at 1,500°C

SiC Fiber

SiC Fiber (Hi-Nicalon)
World production: 3 tons per year

- Diameter : 14µ m
- Tensile strength 3GPa (300kg/mm^2)
- Heat resistance 1700°C
- Density : 2.7g/cm^3

Applications

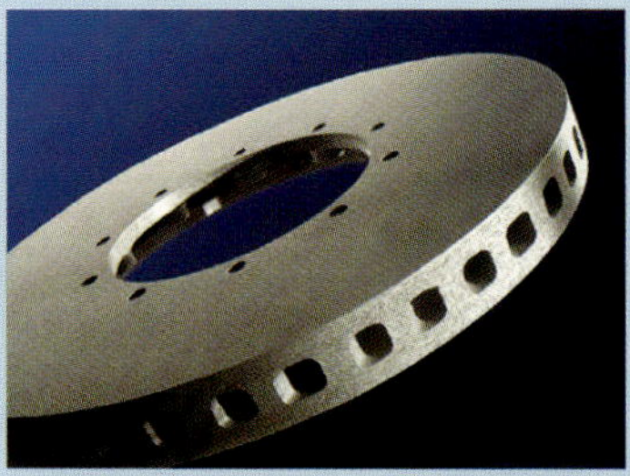

Ceramic Matrix Composites

Space Shuttle

Irradiation

Irradiation makes it possible to create a large number of new materials.

Radiation-induced polymerization can, for example, replace the heating process for fast-drying printing inks, while using a lot less energy. It can also be used for vulcanizing rubber without adding sulfur and for making allergen-free latex gloves, etc.

Radiation-induced cross-linking is the basis of thermoshrinking plastics, which have become key features in electronic connections, and for wrapping products of all shapes and sizes for shipment. It is also used in developing ceramic composites and other advanced materials (see above).

Naval Propulsion

Nuclear-reactor propulsion systems give military nuclear submarines unrivalled autonomy, and as the reactor does not consume any oxygen, they can stay underwater for several months at a time if need be. It could be said that the only real submarines are the nuclear-propelled ones and that all the others are simply submersibles. The United States have also built a fleet of nuclear propelled aircraft carriers, while the French navy also includes one, the Charles de Gaulle (see chapter 4).

Nuclear-reactor propulsion has also been tested for civilian purposes, but the technology has so far only been put to use in one real application, Russian icebreakers. Three nuclear-propelled cargo ships have been built, the U.S.S. Savannah, Germany's Otto Hahn and Japan's Mutsu, but there have been no further commercial applications. Mutsu ran into problems on her maiden voyage in 1974 and had to wait a long time for a port willing to accept the boat to repair a radiation leak. The Russian nuclear-powered icebreaker fleet, on the other hand, has proved to be extremely efficient.

It could be said that the only real submarines are the nuclear-propelled ones and that all the others are simply submersibles

The Yamal, a Russian nuclear-powered icebreaker.

The Mutsu, a Japanese nuclear-powered cargo ship.

Pioneer 10 was launched in March 1972, and its radioisotope generator was still operational in 2003 while the probe was leaving the Solar System!

Space Applications of Nuclear Energy

When a space probe's mission takes it very far from the sun, solar panels are no longer powerful enough to supply the heat and electricity needed, while fuel cells do not have the required autonomy. The only solution is to carry nuclear sources, in the form of radioisotopes or even small reactors. The number of space probes carrying isotope generators is surprisingly high and the illustration below gives only a vague idea. One of the most spectacular success stories was Pioneer 10. The probe was launched in March 1972, and its radioisotope generator was still operational in 2003 while the probe was leaving the Solar System!

With the exception of a single American test launch, only the Russians - or rather the Soviets - have actually used nuclear reactors in Space. During the Cold War, the reactors were used to power imaging radars which consumed too much energy to be powered by solar panels. One of these satellites crashed in northern Canada in 1977 and the political fallout was considerable. The page opposite describes the NERVA program, which was overshadowed by Apollo, but which made a number of far-reaching technical achievements.

Radioisotope Missions

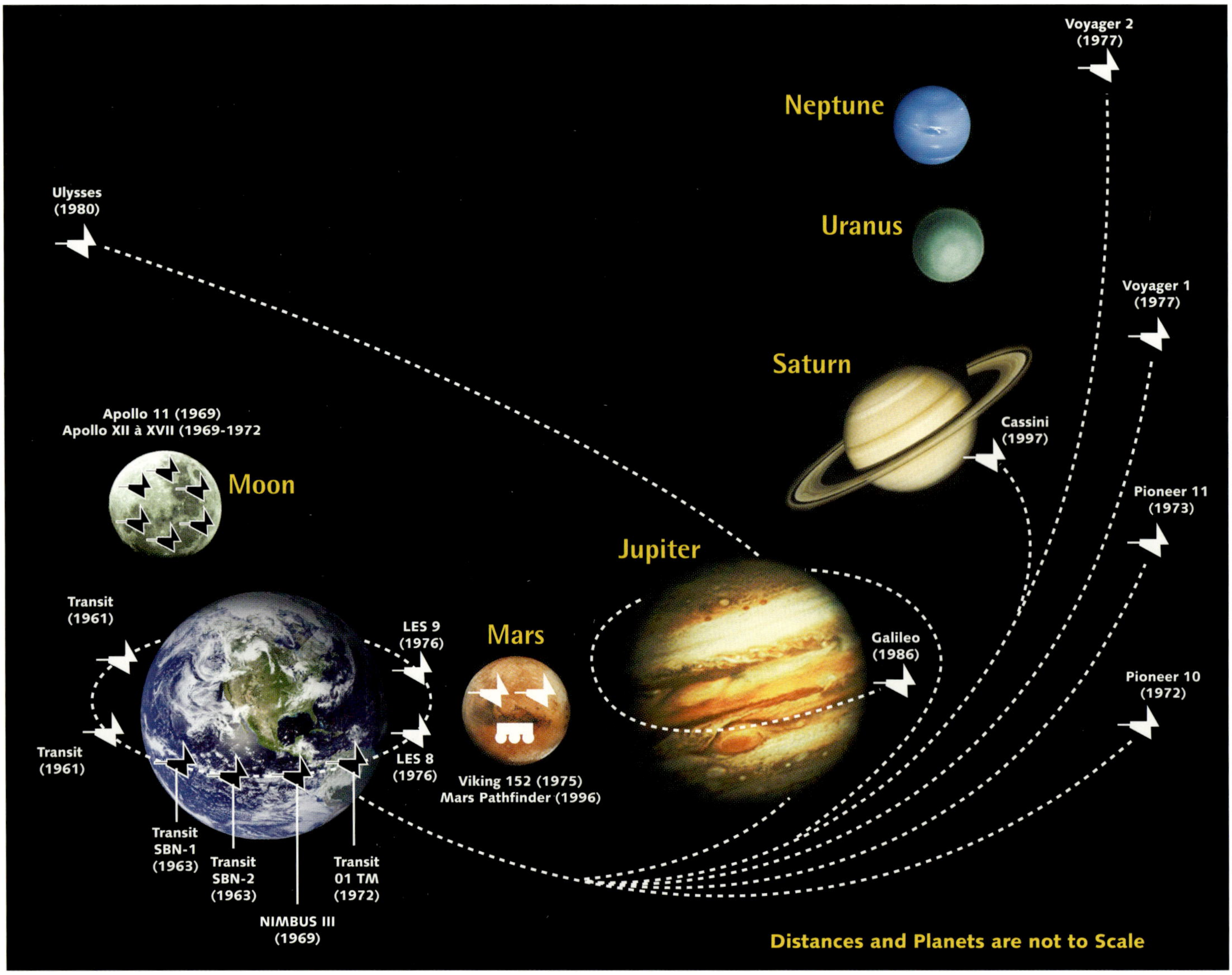

Nuclear Rocket Propulsion: The NERVA Saga

NERVA, Nuclear Engine for Rocket Vehicle Application, was a program carried out parallel to the Apollo Program in the 1960s by NASA and USAEC (U.S. Atomic Energy Commission). The aim was to develop a nuclear engine for space rockets and instead of carrying the huge payload of liquid oxygen and hydrogen needed by a conventional launcher (e.g. the Shuttle), the rocket would only need to carry hydrogen, together with a nuclear reactor for heating it to very high temperatures before ejecting it through the nozzle to create thrust. The NERVA team designed and built a series of reactors that still seem extraordinary some thirty years later.

Despite NERVA's remarkable success, it was abandoned after the American public began losing interest in the space program following the last lunar landing mission, Apollo 17, in 1972, and the general wind-down of NASA in the post-Apollo era.

Design aspects of the NERVA program have recently been revived as part of the project to explore neighbouring planets.

- The reactor was only 1.3 meters long. Hydrogen entered it at a temperature of -180°C and was heated to its exit temperature of 2,500°C
- The temperature increase was higher than 150°C per second
- The ultra-compact reactor-design produced more than 4000 MW per m^3
- Tests on NERVA's NRX series saw the reactor operating for two hours, with three shutdowns and restarts. By way of comparison, an Ariane 5 engine operates only for about 10 minutes...

The temperature increase was higher than 150°C per second

The NERVA NRX at the test site. The nozzle can be seen pointing upwards, while the liquid hydrogen tanks are behind the wall to the left.

A NERVA firing test in the Nevada test site.

A lesser known but very promising application of nuclear techniques is the purification of flue gases

Protecting the Environment

Marking by radioactive tracers has become a basic tool in hydrology, while isotopic analysis is widely use for dating underground aquifers. Tracers are also used to track industrial pollution, including, of course, radioactive pollution.

A lesser known but highly-promising application of nuclear techniques is the purification of flue gases. A mixture of water and flue gas from a coal-fired power plant can be irradiated with an electron beam to eliminate the sulfur and nitrogen oxides that cause acid rain. The by-product obtained from the treatment incidentally serves as an excellent fertilizer!

This process is in use in Japan, China and Poland.

Nuclear techniques have also helped us to understand the environment. The Earth's surface temperature over the last 900,000 years, for example, has been measured through isotopic analysis of ice cores taken from the polar caps of Greenland and Antarctica. For their work on the subject, Jean Jouzel and Claude Lorius were awarded the prestigious gold medal of the French CNRS national research centre.

Fume Purification by Electron Irradiation

SO_2 Elimination : 90% - Absorbed Dose : 10-15Gy - Temperature : 65-70

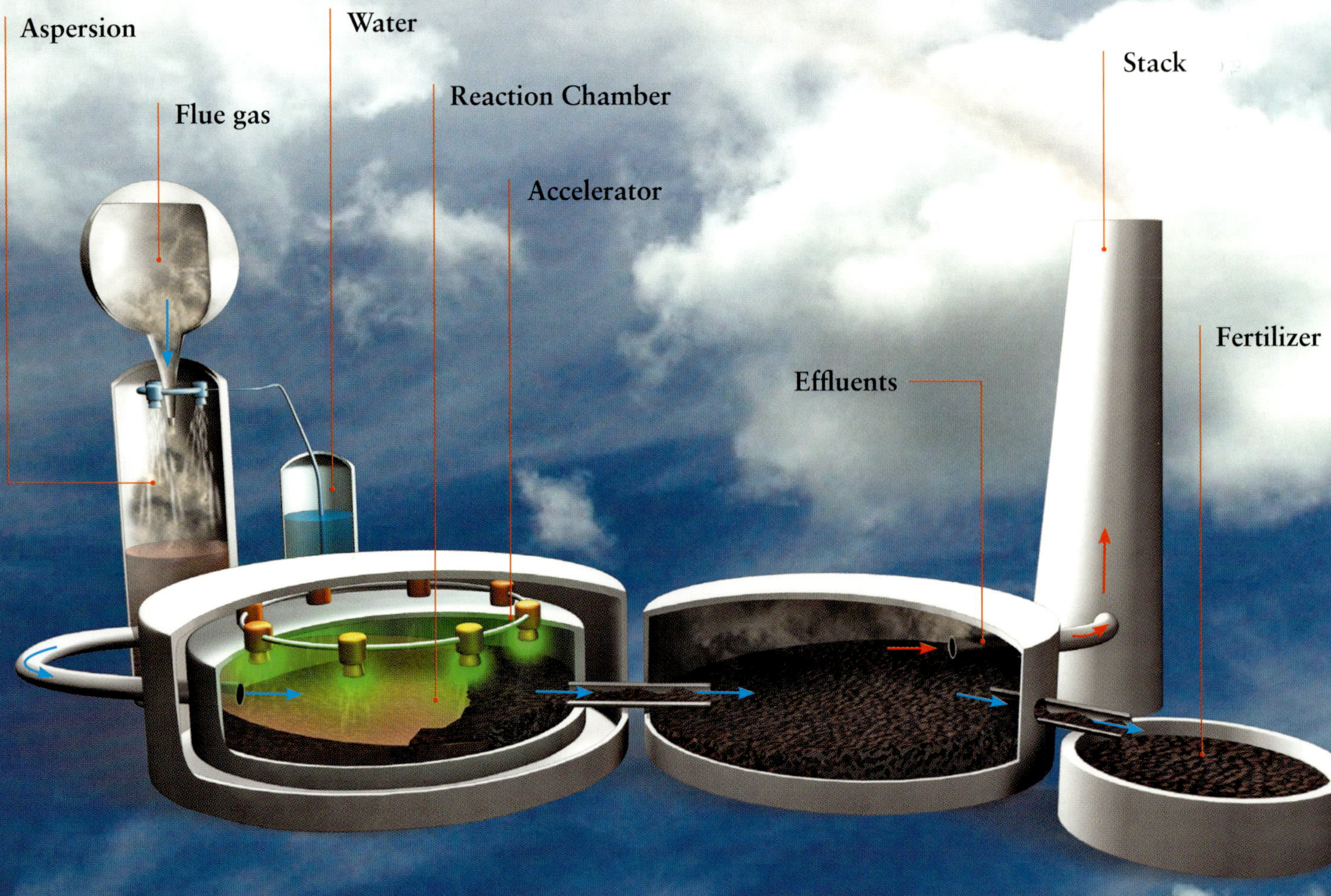

A huge Variety of Applications

We would need far more space than we have here to describe all the peaceful non-power applications of nuclear technologies. Among the huge variety of ways nuclear technology can help us, it is also worth mentioning the combined use of neutrons and photons (produced by synchrotron radiation) for investigating the three-dimensional structure of proteins and other biological molecules, the preservation of ancient wooden statues, right up to the spectacular irradiation of the mummy of pharaoh Ramses II to remove the mould threatening to destroy it.

Radioactive isotopes are also widely used for dating purposes. Carbon 14, for example, is used for dating prehistoric organic remains (wood, charcoal, paintings, artefacts and bones) and for dating minerals to allow us to work out the age of fossils or even mother rocks. This technique is one of the main factors behind the considerable progress made in geological dating over the last half-century.

Scientists and engineers have been investigating other uses of nuclear energy by harnessing its explosive power or using reactors for purposes other than electricity production Some of their ideas have been given full-scale tests, but have been abandoned as they were considered too dangerous or too polluting. These include:

- Civil works applications (digging-dams, artificial lakes, canals)
- Boosting gas production by fracturing oil and gas deposits (widely tested in the former Soviet Union).

There are a great many other ideas that could have a bright future and might well lead to full-scale development, such as the generation of process heat for industrial use and district heating, along with hydrogen production for oil refining and for the chemical industry and possibly also for producing synthetic fuels. Similarly, the use of nuclear reactors for saltwater desalination has already been implemented on a large scale in Kazakhstan. This application could see sharp growth in the near future.

Radioactive isotopes are also widely used for dating purposes. This technique is one of the main factors behind the considerable progress made in geological dating over the last half-century

Irradiation of the mummy of Ramses II, to eradicate moulds which were threatening to destroy it.

Nuclear and Environment

"We must conquer our fears and accept nuclear energy as the one safe and proven energy source that has minimal global consequences."

James Lovelock, The revenge of Gaia, 2006

James Lovelock is the author of more than 200 scientific papers and the originator of the Gaia Hypothesis. He has been described as "one of the environmental movement's most influential figures" (Observer).

People opposed to nuclear power very often invoke the environment as their prime reason for concern, but we are now also seeing the technology's main defenders coming up with the self-same arguments. Does this mean the two sides are moving closer together or are they still barking up different trees?

The environment was not a major issue in the early stages of the development of nuclear power and in any case environmental concerns did not figure prominently in the 1940s and 50s. After the exclusively military focus of the Manhattan project, what grabbed people's attention was especially the prospect of cheap and abundant energy, while another factor was possibly also the moral satisfaction felt by the scientists, especially those who had misgivings about the military use of the atom, at turning swords into ploughshares.

The initial developments, right up until the end of the 50s, showed extraordinary creativity and a boldness, which, looking back, is enough to send a couple of shivers up the spine. The pioneers might have hauled nuclear power out of its infancy, but the USA is now having to clear up the pollution left over from the early days. In the ex-USSR, the situation is even worse, with the pollution legacy in some areas reaching disastrous proportions. Things have, fortunately, changed for the better and the countries which embarked on the nuclear road at a later stage have been able to benefit from the experience of their forerunners and adopt more environmentally-friendly methods. The worldwide nuclear sector is now regulated by very tough rules and standards (see chapter11).

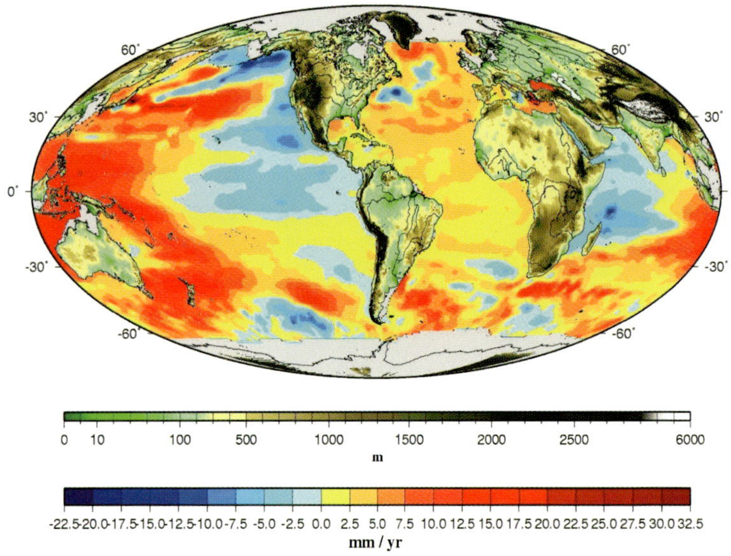

Everybody agrees that nuclear power plants, in normal operation, have a minimal environmental footprint, in other words, they do very little harm to the environment. One of the more surprising facts is that a nuclear power station actually releases less radioactivity into the atmosphere than a coal-fired plant, as coal contains uranium and thorium impurities, which escape through the stack. Nuclear plants also give off none of the atmospheric pollutants, such as sulphur and nitrogen oxides, volatile organic compounds and heavy metals, that are responsible for acid rain. Even the modern "clean-coal" power plants need huge scrubbers to remove the pollutants from the fumes they release.

On the other hand, a major accident like the Chernobyl explosion or the one in 1957 at Kyshtym in the Urals can send huge amounts of radioactivity into the atmosphere. This kind of accident must, therefore be prevented at all costs, and if, despite all the safeguards, one actually occurs, we have to be ready to minimize its impact. This is the aim of nuclear safety and it is what we will be looking at in the next chapter.

Almost all the radioactivity coming from the generation of electricity in a nuclear power plant is contained within the nuclear waste at the back-end of the fuel cycle. We will be examining the management of this waste in the next chapter. The extraction of uranium does not in itself result in any additional radioactivity, but it brings to the surface natural radioactive elements produced during the decay of uranium, some of which are more radioactive than natural uranium itself. In clean-up or remediation operations involving old mines, we therefore have to take great care to isolate all the residuals left over from mining operations and which contain these radioactive elements.

We have looked at the downside of nuclear power, but there is, of course, also a substantial upside (see chapter 14), which is steadily growing in importance. Of all the ways we have of mass producing electricity on demand, nuclear energy releases the least amount of greenhouse-effect gases over the whole cradle-to-grave life cycle. It even emits less than renewable energy sources, which require relatively large quantities of materials for building them.

One school of thought lumps nuclear and fossil-energy sources together in the same basket, arguing that the generation of nuclear waste is just as harmful as the emission of greenhouse-effect gases. The table below gives a few facts which might help everyone decide one way or another.

* PV = photovoltaic,
CCS = Carbon capture and storage

** Ranges reflect differences in methods of assessment, conversion efficiency, scope of assessment, etc.*

Greenhouse-effect Gases	Nuclear Waste
Billions of tons per year	Thousands of tons/year, 95% can be recycled
Freely released	Contained, and stored under surveillance
Global climate threat	Risk of local contamination
Before the end of the century	Not within the next 10,000 years
Modelling forecasts a catastrophe	Modelling forecasts no health effects
Ultimate disposal currently being studied (CCS)	Geological disposal currently being demonstrated
Uncosted	Provisional Cost included (enough ?)

Safety, Radiation Protection and Waste Management

What are the Risks?	134
Defence in Depth	134
The 3 Barriers of the PWR	134
From Design to Operation	136
Safety Organization	136
The Three Mile Island Accident	138
The Chernobyl Accident	138
Protection against Terrorism	139
Radiation Protection	140
The Various Types of Ionizing Radiation	140
Health Effects of Ionizing Radiation	141
Radioactive Waste Management	144

If accidents are to be avoided, the first thing to do is to identify any failures that can cause them. The plant has to be examined to find ways of protecting it from any such dangers and where needed, to install systems capable of strengthening its lines of defense

The nuclear industry handles large quantities of potentially hazardous radioactive materials. It can as such be compared to those other branches of industry which work with often highly toxic products, which, like radioactivity, can also be colourless and odourless, but which are much less easy to detect and measure.

The overriding safety objective in nuclear facilities is to guarantee under any circumstances the containment of radioactivity.

What are the Risks?

Nuclear facilities face the same conventional risks as many other industrial plants, including fire risks, falling loads and pressurized steam. These risks can be harmful for the employees but can also cause damage to those very systems that are designed to contain radioactivity. Extremely tight controls are therefore in force in nuclear facilities to prevent these risks happening.

Nuclear facilities are also highly protected against external risks, whether they result from human activities (plane crashes, shock waves from off-site explosions and acts of terrorism), or from natural phenomena (floods, earthquakes, storms, etc.).

There are two other risks specific to the nuclear industry that need to be kept under tight check:

The first risk comes from the chain reaction, an inherent part of nuclear fission. The reaction has to remain fully controlled in nuclear reactors, and cannot be allowed to occur at all in other types of nuclear facility. This is called the criticality risk.

The second specific risk comes from the heat released by radioactive decay, suitably enough called decay heat. Ongoing removal of this heat is a major safety concern, whenever significant quantities of radioactive materials are to be found together. This is especially true for reactors, which need extremely reliable cooling systems which will continue to operate after shutdown. Safety in nuclear reactors is based on the three following criteria:

- Controlling the chain reaction and therefore the power generated;
- Cooling down the fuel, even after shutdown, to evacuate the decay heat;
- Containing the radioactive products.

Safety in the other types of nuclear facility is like in the fuel cycle, and primarily involves containing the radioactive products, while minimizing the probability of a nuclear criticality accident.

Defense in Depth

A plant will be safe as long as it can put up sufficiently strong lines of defense against any potential failure or danger. The strength of the defenses does not just depend on how good they are but also on how they are massed. The basic premise is that any safety system, no matter how reliable it is, is vulnerable and must be reinforced or protected by a back-up system. Similarly, each line of defense both reduces the probability of an accident occurring and lessens the possible effects of any such occurrence. Massed lines of defense bring risk levels down so low as to be deemed acceptable

The defense-in-depth approach, whereby the plant is protected by successive lines of defense, provides what has generally proved to be an effective cornerstone for nuclear safety.

The 3 Barriers of PWR

First and foremost, a reactor has to be well designed. One simple but highly-efficient design feature involves setting up a series of three extremely tough, leaktight barriers between the radioactive products and the environment to form a triple protective shield to prevent radiation escaping and to contain the radioactivity under any circumstances.

• The ceramic fuel pellet, which already contains most of the radioactive products, is enclosed in leaktight metal cladding. This is the first barrier.

• A second barrier is formed by the thick steel walls of the primary coolant circuit, which includes the reactor pressure vessel containing the reactor core.

• Finally, the entire primary system (as well as other parts of the reactor) is contained within a solid concrete shell capable of withstanding high pressure and external hazards or attacks. This is the containment building. It is also completely leaktight and often double-walled.

While any one of the three barriers can on its own effectively contain radioactivity, the main objective must, of course, be to preserve the integrity of each of the three under any circumstances. Reactor design therefore includes three types of system, used in different combinations.

The first system is based on natural physical phenomena and no human intervention is required. Reactors are designed so that any core temperature increase caused by excess power spontaneously stops the chain reaction. Similarly, any significant loss of water in the core automatically quenches the reaction.

The second method uses passive safety systems. One example is the neutron-absorbing control rods, which, in the event of a power failure, automatically fall into the core to stop the reaction, while another system automatically injects water into the core from accumulators. Control rods are under the control of the plant protection system and automatically shut down the reactor as soon as plant operating conditions move outside acceptable norms.

The third method uses active safety systems powered by electricity or pressurized air, and triggered by signals from sensors. Any water loss in the core, for example will trigger high pressure pumps (later replaced by passive accumulators).

On top of these already extensive safety measures, the systems are diversified and given a double, triple or even quadruple back-up to make sure that no single event, such as a fire or flooding, can simultaneously disable several systems performing the same function (common mode failures).

From normal operation to accident management

First level:
The lines of defense (safe design, high quality construction and tightly-controlled operation) ensure normal operation.

Second level:
Surveillance devices to detect any anomaly and redundant protection systems to bring the reactor back to a safe state.

Third level:
If the first two levels are overcome, a set of redundant safeguard systems is included to make sure that one barrier at least remains intact.

Beyond:
Emergency Planning and Procedures

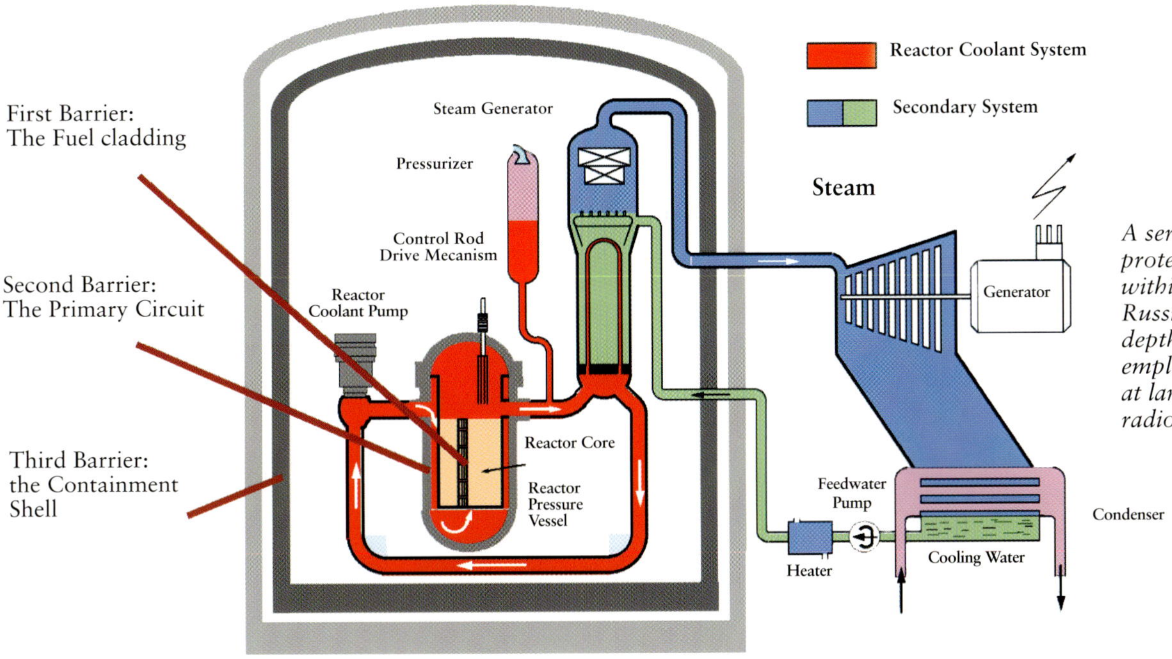

A series of independent protective barriers, encased within each another like Russian dolls, provide in-depth protection for both employees and the public at large against nuclear fuel radioactivity.

The INES International Nuclear Event Scale was designed along the same lines as the Richter earthquake scale, and the Beaufort wind scale, with the aim of providing the media and the public with a way of assessing the severity of an event (failure, incident or accident) affecting a nuclear facility. Since 1991, events have been graded on a sliding scale of increasing severity from 0 to 7. The only level 7 event recorded to date is the Chernobyl accident.

In France, any events involving nuclear safety must be declared by the operators to the Safety Authority (ASN) within 24 hours, and the operators gives their INES assessment of the severity of the event. The final INES rating will be decided by the ASN. Each year French operators declare several hundred level 0 events (which have no safety significance), about one hundred level 1 events and between one and three level 2 events. There were two level 3 incidents in 1981 and in 2002, and one level 4 in 1980.

From Design to Operation

A safe reactor is not just one with a good design, it also has to be actually built in strict compliance with the specifications laid down during the design phase. A quality assurance program is maintained and applied throughout the whole of the design and construction phases of the reactors in order to ensure such compliance. High construction quality is therefore the second line of defense.

The third and final line of defense is tightly-controlled, highly disciplined operation by trained specialists working with clear, precise procedures and having full-time, instant access to all the information needed to drive the plant.

As an additional safety feature designed to improve the systems described above, a system has been set up in all nuclear power plants to gather data on any operating events (failures, malfunctions, and incidents) in order to improve all the above-mentioned measures. The data are analyzed on an ongoing basis to learn from experience and make any necessary changes to prevent the recurrence of such events. This process is called Return of Experience.

The safety measures and systems currently applied in nuclear reactors mean that the risk of a core meltdown in a fleet of 1000 PWRs has been estimated at lower than 1 for every hundred years of operation. This kind of accident has, however, already occurred, when the Three Mile Island nuclear power plant, in Pennsylvania (USA) suffered a partial meltdown in 1979. Fortunately, there were no casualties and the containment building was able to maintain its integrity, thereby preventing serious damage to the environment. The Three Mile Island accident was graded level 5 on the INES scale (see below).

Safety Organization

Nuclear safety includes all the systems implemented at every stage of the design, construction, operation, and decommissioning of both nuclear facilities and radioactive material transports both to prevent accidents and contain their effects, should such an accident occur.

Safety comes under the sole responsibility of the facility operators, but is supervised by the public authorities, whose duty it is to ensure that this responsibility is fully met and is compliant with the regulatory framework.

CRITERIA OR SAFETY ATTRIBUTES	OFF-SITE IMPACT	ON-SITE IMPACT	DEFENCE IN DEPTH DEGRADATION
7 MAJOR ACCIDENT	Major release: widespread health And environmental effects		
6 serious ACCIDENT	Significant release likely to require full implementation of planned counter measures		
5 ACCIDENT	Limited release: likely to require full implemtation of planned counter measures	Severe damage to reactor core radiological barriers	
4 ACCIDENT	Minor release: public exposure of the order of prescribed limits	Significant damage to reactor core/ radiological barriers/fatal exposure of a worker	
3 SERIOUS INCIDENT	Very slight releasepublic exposure at a Fraction of prescribed limits	Severe spread of contamination acute health effects to a worker	Near accidents-no safety barriers remaining
2 INCIDENT		Significant spread of contamination over eposure of a worker	Incidents with significant failures In safety provisions
1 ANOMALIE			Anomaly beyhond the authorized operating conditions
0 DEVIATION		No safety significance	
OUT OF SCALE EVENT		No safety relevance	

Operators will, of course, usually set up their own internal control systems as part of their responsibilities, but independent control is an indispensable factor in nuclear safety.

The respective roles of public authorities and the operators can be summarized as follows:

- Public authorities lay down the general safety objectives.
- Operators come up with the technical solutions designed to meet these objectives and are subsequently required to demonstrate their efficiency
- The public authorities ensure that the operators' proposed solutions are compliant with the safety objectives and check that the safety demonstrations are also in line with these objectives.
- Once approved, the operators implement the solutions.
- The public authorities inspect the facilities to ensure that the solutions have been properly implemented, and take action when necessary.

In democratic countries, there is also an additional layer of citizen control, mainly through Parliament but also through organizations where all the stakeholders are represented (trade unions, associations, media, academics, etc.). The role of this citizen control is to assess overall safety system efficiency and to ensure transparency in nuclear operations.

In France, safety organization was set out in an Act, dated June 13, 2006 which related to the "transparency and safety in the nuclear sector", which now comes under an independent administrative authority, the ASN, the nuclear safety authority, which works alongside independent expert bodies. ASN is a similar type of organization to America's USNRC or Britain's NII. Local information commissions have also been set up in France to provide the public with information relating to the nuclear industry.

Three Mile Island

Aerial view of the Three Mile Island NPP (Pennsylvania). The cooling towers of unit 1, still operating, can be seen at the top half of the picture. Unit 2, below, was decommissioned after the accident of March 29, 1979.

The Three Mile Island Accident

On March 29, 1979, on Unit 2 of the Three Mile Island (Pennsylvania) PWR, pumps supplying feedwater to the steam generators failed, triggering an automatic shutdown. A pressurizer relief-valve failed to close and the emergency system then activated the high-pressure safety injection system. Misled by an ambiguous signal in the control room, the operators thought the reactor was overfilling and they manually shut off the water injection. Two hours later, when they realized there was a leak, they restarted the injection, but by then a third of the core had melted and the rest was severely damaged. The news of the accident terrified the USA and the reactor was damaged beyond repair. Fortunately, the containment building maintained its integrity and, apart from the release of relatively innocuous radioactive rare gases, it prevented radioactivity from the destroyed core from escaping. No deaths or injuries were recorded in the neighbouring populations.

Defense-in-depth was seen to work - the plant design was safe, construction was sound and the protective and safety systems operated as they were supposed to. Human error over a fairly extended period caused the core to melt, but the third barrier held firm.

The Chernobyl Accident

The Chernobyl (Ukraine) nuclear plant consisted of reactors based not on the PWR design but on the RBMK, a type built exclusively in the former Soviet Union. In RBMKs, the fuel assemblies are cooled by ordinary boiling water flowing under pressure through vertical pressure tubes which run across a huge graphite pile. The reactors used graphite (and not water) to moderate the neutrons. RBMKs have a number of design flaws: the reactors are unstable at low power, they have a dangerously large positive void coefficient (the chain reaction accelerates if the water density in the core decreases) and their third barrier is not complete, offering only partial containment

On April 28, 1986, the crew of Chernobyl Unit 4 were carrying out an experiment on the reactor at reduced power, which required them to disable some of the plant's protection systems. After a time, the reactor became unstable, but the crew chose to continue the experiment and not shut it down and further exacerbated the situation by violating a series of safety procedures. Reactor power began to rise very rapidly and the core cooling water was turned instantly into steam, which under increasing pressure, caused a steam explosion that blew the 1200 ton reactor lid off the top of the core. The lightweight roof of the reactor building was blown away and part of the core itself was exposed. The rush of oxygen in the open air caused the graphite to overheat and catch fire. The fires lasted several days, and sent a huge amount of radioactive aerosols high up into the stratosphere later, to fall all over Europe. The different effects of Three Mile Island and Chernobyl dramatically underlined the importance of the containment building.

Even the best-designed and best-built reactors in the world would not be safe in the hands of untrained or unskilled operators

Protection against Terrorism

When safety is involved, the need for transparency is paramount (except for details which are covered by commercial or intellectual property rights). Official Secrets as such only apply to the physical protection of nuclear materials. However, as far as protection against acts of terrorism is concerned, secrecy understandably becomes a matter of importance and transparency has to take a rear seat..

The efficiency of the steps taken to protect facilities from terrorist or other criminal acts is partly based on the fact that would-be perpetrators do not know the details of the systems implemented. Banks, for example, are not going to give away the combination of their vaults. This being said, a lot of the facts and features are in the public domain or can be found out with a quick search on the Internet or elsewhere.

With all their safety systems and procedures and the special design features for protecting them against terrorist attacks and natural disasters such as earthquakes, storms and floods, it is hardly surprising that nuclear reactors are among the mostly solidly-constructed and least vulnerable industrial facilities outside the military field.

This does not mean a nuclear reactor would come out undamaged if an airliner crashed or was crashed into it. There would be damage and quite some time would be needed before it was repaired and authorization granted by the Safety Authorities to resume operation.

The risk, however, that such a crash would cause significant release of radioactivity in the environment is minimal. The main danger would be the ensuing panic it might cause both among and beyond the local population.

The EPR reactor is designed to afford maximum protection in the event of external attack or natural disaster. It features quadruple protection and safety systems, located in 4 different buildings so that they cannot be all put out of service at the same time.

World Trade Center 2001.

The buildings are either protected by a thick reinforced concrete shell or located separately on opposite sides of the containment building.

Protection against terrorism does not belong to nuclear safety as such, but is part of wider security issues covering sensitive facilities. The September 11, 2001 terrorist attacks against the New York World Trade Center have caused a massive increase in anti-terrorist concerns

Radioactivity is a Natural Phenomenon

When the Earth was formed some 5 billion years ago, it consisted of a mixture of stable and radioactive elements. Since then, Earth's radioactivity has been going down all the time, as most of the radioactive atoms have turned into stable nuclei.

Radiation Protection

Ever since the world began, every living organism has been constantly exposed to radioactivity. Nuclear radiation is all around us in the environment and we breathe in and eat radioactive elements all the time. But before going any further, let's have a recap of the units of measurement

• Radioactivity is measured in becquerels (Bq): 1 becquerel is equivalent to one nuclear decay or other nuclear transformation per second.

• One gray (Gy) is the absorption of one joule of radiation energy by one kilogram of matter.

• The sievert (Sv) measures the biological effects of radiation absorbed by living tissue, according to the type of radiation and the organs exposed. The result is called the equivalent dose and is often measured in millisieverts, mSv. 70% of the radioactivity we are exposed to is natural and comes from 4 main sources:

• Cosmic radiation from the Sun and space. This varies with latitude and especially altitude, rising from 0.4 mSv per person/year at sea level, to 1.7 mSv at an altitude of 4,000 m.

• Terrestrial radiation emitted by radioactive elements found at varying concentrations in the Earth's crust. Exposure ranges from 0.4 mSv per person/year in the Paris Basin, to 1 mSv in Brittany or the Massif Central, to more than 10 mSv in some regions of Brazil, India and Iran.

• The air we breathe contains radon, and its decay products are deposited in the respiratory tract. The mean dose equivalent is around 1.3 mSv per person per year and varies according to a number of factors such as ventilation and the type of construction materials

• Food and drink contain radioactive elements that are deposited in tissue and bone. The human body contains on average 4,500 Bq of potassium 40 and 3,700 Bq of carbon 14, amounting to a dose equivalent of 0.2 mSv

Some examples of radioactivity levels:

• Rainwater: 0.3 to 1 Bq/liter
• Mineral water: 6 Bq/liter
• Seawater: 13 Bq/liter
• Milk: 80 Bq/liter
• Human Body: 100 Bq/kg
• Granite: > 1,000 Bq/kg

Worldwide, the dose equivalent due to natural exposure varies considerably but the average is around 2.4 millisieverts per person/year. 30% of the radioactivity we are exposed to comes from industrial radiation, and especially from medical uses. By way of comparison, the nuclear energy industry as a whole accounts for under 0.02 mSv per person/year.

Different Types of Ionizing Radiation

Radiation is said to be ionizing when it has enough energy to remove electrons from atoms or molecules in the irradiated medium.

Ionizing radiation includes:

• Very high-energy cosmic rays.

• The highest-energy electromagnetic waves: X and γ radiation, which lose little energy upon contact with matter and can penetrate quite deeply.

• Charged particle flows: α radiation (helium nuclei emitted), β radiation (neutrons transformed into protons in the nucleus and electrons emitted), and β+ radiation (protons transformed into neutrons in the nucleus and positrons emitted).

• Neutron beams.

Different Exposure Pathways

Ionizing radiation can reach an organism via external sources, upon contact with the skin, or from within inside the organism. The first case is called external exposure, the second external contamination and the third internal irradiation.

External Exposure

This results from exposure to a radiation field produced either by a generator of electromagnetic radiation or by an electron generator. For a given radiation source, three factors determine the level of exposure to external irradiation: the exposure time, the distance from the source and the presence of any shields between the source and the person.

External Contamination:

Radioelements that get on the surface of the skin irradiate it with the ionizing radiation they emit. This type of contamination can be treated by thorough washing and checking for any residual activity.

Internal Irradiation

The irradiation is the result of internal contamination, meaning that radioelements have entered the organism through the respiratory or digestive tracts, through injuries or wounds, or through the skin. Treatment involves the use of decorporating chelating agents to remove the radioactive elements from the various organs. Another

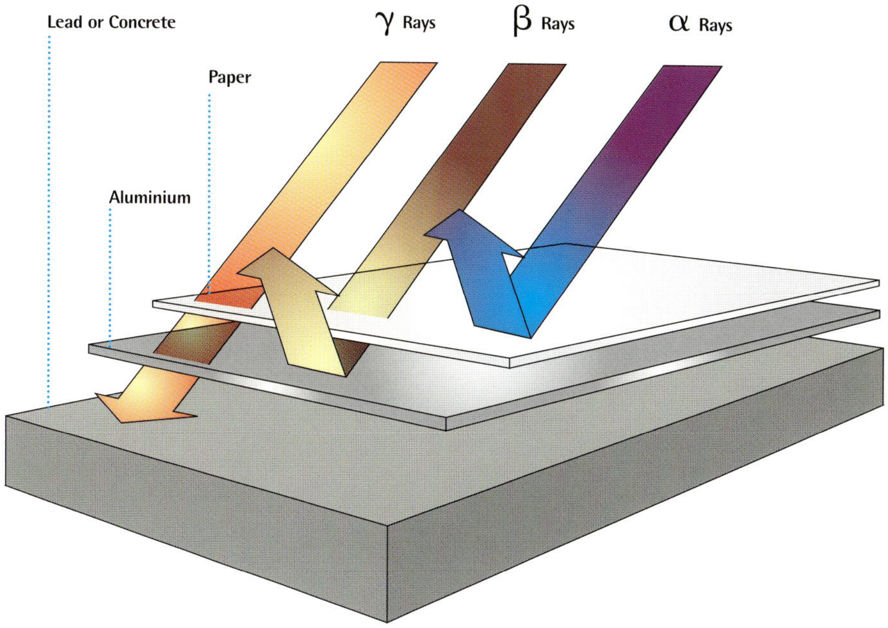

approach is to prevent internal irradiation by saturating the organism with the corresponding stable element. This method is used for iodine.

Ionizing Radiation and its Effects on Health

The biological effects of radiation have been common knowledge since a few years after radioactivity was discovered, while the dangers of radiation were known long before it was put to industrial use. The dangers can therefore easily be factored into the design phase of facilities.

Two methods are currently used to study the various biological effects of ionizing radiation on health: the first is epidemiology and the second radiobiology, which involves carrying out experiments on molecules or cells from living tissues:

• The epidemiological approach consists of studying the effects on an irradiated population (Hiroshima and Nagasaki survivors, pioneering radiologists, etc.) and comparing them with a non-irradiated population to establish dose/effect correlations. The lower the dose, the larger the cohort population (the people being observed in the experiment) needs to be for the results to be significant.

A lot of the over-the-top figures found in the press and in books, even the most scholarly ones, about the number of excess deaths caused by the radioactivity released in the environment following the Chernobyl accident come from inaccurate use of the LNT (Linear No Threshold) model, and its use to assess the effects of very low doses of radiation on large populations.

The study of the health effects of radiation borrows from both physics and biology. Although there has been considerable progress, there still remain gaps in our knowledge. If we want to get a precise picture of the effects of radiation, we need to find out the radiation dose received by every part of the organism, which is not currently within the realms of possibility.

• Researchers have been able to carry out experiments to observe the effects of ionizing radiation on animal and cell cultures and gain insight into the intimate mechanisms of lesions. It is difficult, however, to apply the results of studies on DNA to the effects on cells and even more difficult to apply results from cell studies to the effects on tissue.

Seen from the cellular physiology angle, ionizing radiation energy creates lesions in molecules, especially DNA, the very long "double helix" biological molecule which carries the genes.

It should be emphasized, however, that DNA lesions are not specifically caused by ionizing radiation, as lesions are occurring all the time (and are being repaired), induced by free radicals produced within our cells by our own metabolic processes, such as breathing and digestion. Our metabolism causes some 30,000 lesions to be produced every day in every single cell in our body, while natural radioactivity causes about 2 per year! This provides an insight into the incredible efficiency of our body's repair system. Certain types of radiation which deposit energy over short distances (neutrons) can, however, cause simultaneous, adjacent lesions on both DNA strands, (the so-called double-strand break) and repair then becomes much more of a problem. Damage to a DNA strand triggers these highly-efficient repair mechanisms. If the repair is successful, nothing else happens. If, however, the repair fails and the defect is detected, it results in cell suicide (a mechanism known as apoptosis), where no harmful substances are released. If the defect goes undetected, which fortunately is very rare, then there is a risk of the cells becoming cancerous.

Dose-Effect Relationship

The relationship between the intensity of a radiation dose and its effect on health effects can be studied in two ways:

• Deterministic: a causal relationship is sought between dose and effects. This approach, which can only be adopted for high doses and acute effects, produces a threshold effect: a total body dose of 5 grays causes near-certain death but, conversely, no symptoms at all appear below the 0.25 Gy threshold

• Stochastic (or random): when a population is irradiated at lower doses, delayed effects appear in some individuals, but their causes cannot be determined for certain. In such cases, the effects recorded with high doses are extrapolated using an LNT (Linear No Threshold) model. This method is highly conservative (and somewhat disputed), but is widely used to set, for example, maximum acceptable radiation exposures. It should not, however, be used for assessing the effects of collective doses (see inset on next page).

These are the (deterministic) effects of doses absorbed by single full-body irradiation:

• Below 250 mGy: no observed effect
• From 250 to 2,500 mGy: symptoms appear but recovery is almost certain
• Between 2.5 and 5 Gy: serious damage to health.
• Above 5 Gy: death is almost certain.

International Radiation Protection Norms

The potential danger of excessive exposure to ionizing radiation led to the establishment of international regulatory standards setting the acceptable radiation dose levels. These levels refer to a minimal excess risk compared to the risk from normal exposure. The International Commission on Radiological Protection, founded in 1928, puts forward recommendations on the principles of radiation protection that form the basis for codes and regulations issued by other international organisations and regional and national authorities. The European Union has incorporated the ICRP's recommendations in its own directives.

Current ICRP recommendations quote the following maximum exposures (ICRP 60):

• 20 mSv per year, averaged over 5 years (with a yearly maximum of 50 mSv) for workers in the nuclear sector.
• 1 mSv per person per year for the rest of the population

The effects of radiation can be beneficial when directed against abnormal cells or harmful if used on healthy cells. The effects vary widely as radiation can affect a simple water molecule just as much as a fragment of DNA. Living tissues have natural repair mechanisms that are highly effective if the radiation dose is low.

The Linear no Threshold Model: Recorded Deaths, computed Deaths, virtual Deaths?

The use of the LNT relationship to determine the maximum acceptable radiation exposures for the protection of workers and the public in general is a perfect illustration of the "precautionary principle", which the French have made part of their Constitution. LNT is much less credible, however, when used to assess the effects of collective doses. To take a somewhat extreme example of the misuse of the LNT model:. 6.5 billion human beings are exposed annually to an average natural radioactive dose of around 3 millisieverts. The collective dose is therefore 19.5 million sieverts, 4 million times higher than the dose which would cause certain and rapid death through radiation sickness.

If 4 million people every year are being killed by radiation sickness resulting from natural radiation, we can be fairly certain that someone, somewhere, is going to notice. Radiation sickness is a deterministic effect , which everybody agrees has a threshold. However, when it comes to assessing stochastic effects such as excess cancers, LNT needs to be used with great care. A lot of the over-the-top figures found in the press and in books, even the most scholarly ones, about the number of excess deaths caused by the radioactivity released in the environment following the Chernobyl accident come from inaccurate use of LNT and its use in assessing the effects of very low doses of radiation on millions of people.

That is why the Chernobyl Forum chose to restrict its 2005 assessment to the 600,000 individuals who were exposed to radiation over (sometimes considerably so) the natural dose level. The Forum put the total number of premature deaths caused by the radiation released during the Chernobyl accident at a maximum of 4,000. That of course is 4,000 too many, but it is a far cry from commonly quoted figures. In truth, we shall never know for sure, as the figure of 4,000 is lower than the statistical margin of error for the number of "natural" cancers which each year are responsible for a quarter of all deaths in the world.

The Chernobyl accident was a genuine catastrophe, and the only one so far to have been caused by nuclear electricity production, but its effects have been greatly exaggerated by the media.

In all the polls about nuclear power, answers like: "I'm not really against nuclear power...but there is the waste problem!" frequently come up

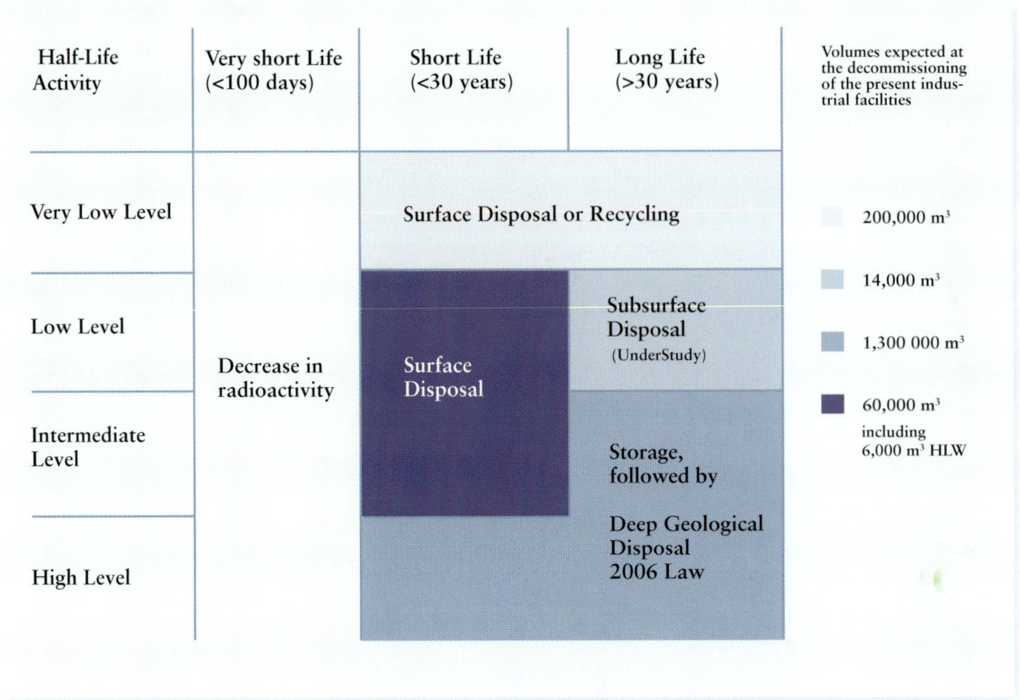

Storage centers in France: in the foreground, the very low level waste disposal center and in the background the short-lived low and intermediate level waste disposal center.

It is worth noting that medical exposure to radiation is not included in the figures, as it is meant to be a beneficial use of radiation, and therefore not counted. Also not counted is exposure to natural radiation, as if it was, we would have to get everyone off the planet as quickly as possible.

Radiation protection also involves applying three fundamental principles and these are as follows:

- Justification of a practice: no person should be exposed to radiation unless the exposure produces more good to that individual or to society than harm.
- Optimization of protection: being below the limit cannot be considered enough in itself, exposure doses must be kept as low as reasonably achievable (ALARA principle),
- Individual dose and risk limits: dose limits should be based on the exposure of individuals to accumulated radiation resulting from a combination of practices.

Radioactive Waste Management

When opinion polls ask questions about nuclear power, one of the answers that always seems to come up is "I'm not really against nuclear power… but there is the waste problem!" When the questions dig deeper, people seem to have only a fuzzy notion of a problem that appears both threatening and unsolvable. In most cases, they do not seem to have any idea about what or how dangerous this nuclear waste they are so worried about actually is!

The French government became so concerned about the issue that they incorporated the following definitions into the 2006 radioactive materials and waste management law.

"A radioactive substance is a substance which contains natural or artificial radioactive nuclei in amounts or concentrations high enough to warrant radiation protection measures."

"A radioactive material is a radioactive substance which has a foreseen use, following processing if needed."

Disposal center for short-lived low and intermediate level waste. Once filled up, the concrete cells are covered with a clay layer.

"A nuclear fuel assembly is considered spent when, following irradiation in a reactor core, it is permanently removed."

"Radioactive waste is a radioactive substance which has no foreseen use in the present technical and economic context."

"Ultimate radioactive waste consists of radioactive waste which can no longer be processed under the prevailing economical and technical conditions, either to extract recyclable materials or to make it less polluting or dangerous"

Legally, therefore, the notion of waste has nothing to do with Nature but has to be taken within the context of foreseen future use. Activity and concentration, through which a radioactive substance is defined, are both norms which are laid down in law.

Nuclear facilities are authorized to release a certain amount of radioelements into the environment, in liquid or gaseous form, up to a maximum annual limit, which is set out by law. As the low concentration of these radioelements does not warrant any radiation protection measures, they are not considered as radioactive waste but as releases or effluents. Unlike releases and effluents, radioactive waste is more often than not solid.

Inside view of a very low-level waste disposal cell.

Internationally, radioactive waste is often divided in three categories:

- "A" waste, of low activity and containing very few elements with a half-life of over 30 years;
- "B" waste, of low and intermediate activity, but containing long-lived elements;
- "C" waste, of high activity.

A waste comes from the nuclear industry, laboratories, hospitals and various industries, while B & C waste comes only from the nuclear industry (spent fuel or ultimate waste produced during reprocessing).

ANDRA

Vitrified waste: looks exactly like volcanic glass (Obsidian) and presents very similar caracteristics.

Partitioning + Transmutation minimize long Term Radiotoxicity

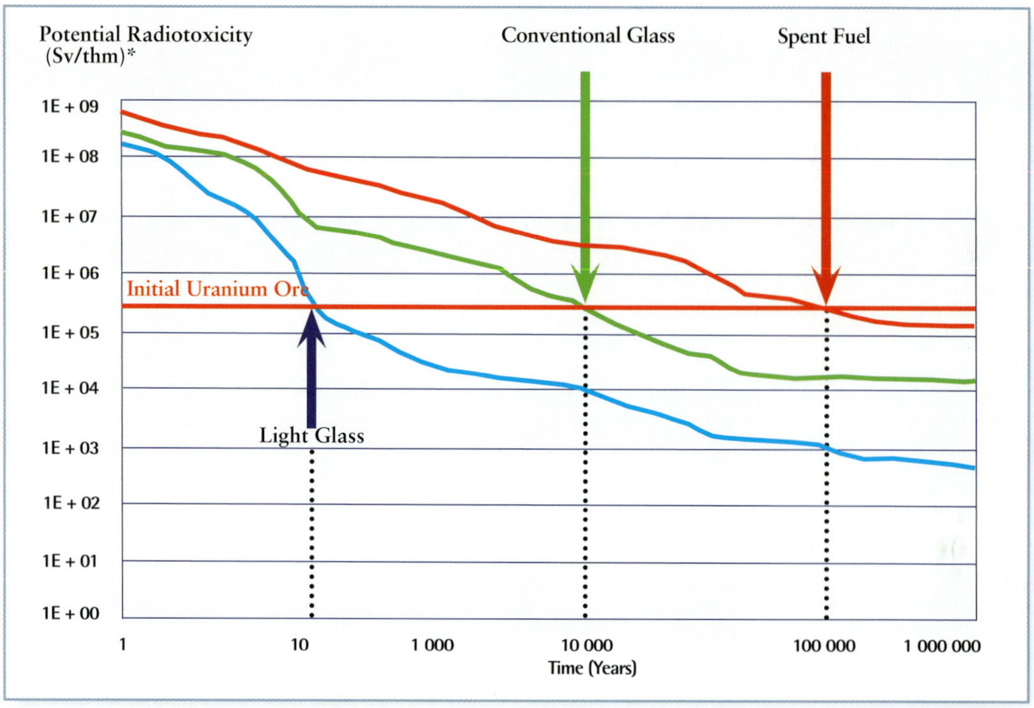

* *Tons of heavy metals*

Sandia National Laboratories

The table on page 144 shows the more detailed classification used in France, together with the quantities expected to be produced by the French nuclear programme until the currently operating facilities are decommissioned

Containment, a basic Principle

Despite the fears it raises, radioactivity has two major qualities:

• It can be detected at levels far below those of any toxic or noxious substance (it is possible to detect radiation from a single decay while for a chemical substance a few billion billion molecules are needed for it to be detected).

• Once radiation has been detected, it is very easy to protect yourself it by keeping a suitable distance from the source, limiting exposure time and/or using adequate shielding.

Final waste disposal therefore comes down to a question of containment - making sure that the radioactive elements present in the waste package will stay put, or that any migration of radioactive materials to the human environment takes long enough for their activity to have decayed to such a degree as to be well within current safety limits.

Waste vitrification.

Low-level (LLW) and very low-level (VLLW) waste

VLLW and LLW represent large volumes, but only contain a small amount of radioactivity. Very low-level Waste is still mostly a French notion and consists mainly of rubble from the dismantling of nuclear facilities, and is regarded more as suspect than actually radioactive. It is safely disposed of in a suitable Center.

Many countries have Centers that have been authorized and licensed by the national safety authorities to dispose of short-lived low and intermediate-level waste. The waste is stored in concrete and covered by earth to a depth of several meters. It will remain under constant supervision for two or three hundred years until its radioactivity has decayed to the same level as the natural radioactivity of its immediate environment (for every 10 half-lives, the ac-

Transportation casks for long-lived waste to be disposed of at WIPP (USA).

tivity of a radioactive element is divided by a factor of 1000). No particular concerns have been raised in these countries about this category of waste.

High-level HLW and long-lived inter mediate-level LL-ILW Waste: storage, disposal, transmutation

Intermediate-level waste containing long-lived elements (LL-ILW) and high-level waste (HLW) are the most difficult to dispose of and they need to be stored in sites which can provide secure containment over periods of several tens of thousands of years, well beyond the storage times for other forms of waste.

Containment of intermediate and high-level waste presents the same kinds of difficulty as containing the radioactivity within a reactor, but with two main differences: the activity concentration (the quotient of the activity of a material and the volume of this material) of HLW is far smaller than that of a reactor core, but the containment period is far longer.

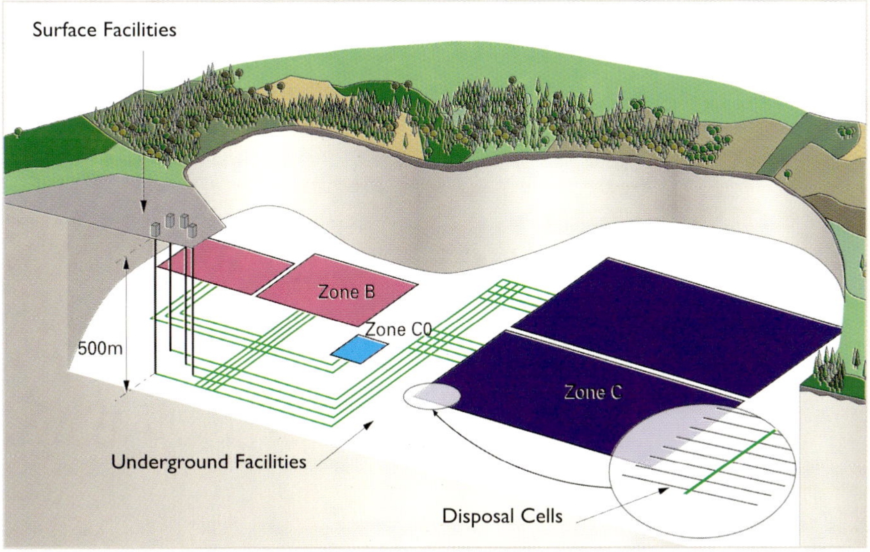

On March 26, 1999, the first long-lived waste package resulting from American Defense programmes was hauled down into the WIPP, Waste Isolation Pilot Plant located in an underground salt layer in the New Mexico desert, near Carlsbad. WIPP was the first LL-ILW deep geological disposal site in the world

The solution for long-lived waste is the time-tested method of multiple nested confinement barriers. The first barrier is the matrix within which the radioelements are immobilized, the second is the waste packaging, and then additional barriers are added according to the method chosen for final disposal.

The basic alternative is between long term storage in a dedicated surface facility (or "subsurface", if the facility is covered by earth down to a depth of several meters) and deep geological disposal. A future solution, to be used alongside either of these two methods, is the transmutation of long-lived radionuclides.

In the long-term storage method, conditioned waste packages are placed in a specially designed facility for a set length of time and at the end of this period - which may be very long - the packages will be retrieved and shipped away from the facility. The storage building can be built on the ground or be buried at shallow depth to provide better protection against potential intrusion or attack.

Both surface and subsurface storage facilities must be kept under permanent, well-guarded surveillance over the whole of the specified period, and the packages must be designed so that they can be retrieved when the decision is taken to do so. Long-term storage is a satisfactory solution for the medium term, but is not a permanent one.

In the deep geological disposal method, the geological stratum itself constitutes the ultimate barrier against the migration of radioactive elements. Once full, the underground facility will be sealed and there will be no future retrieval of the packages. This, at least, was the idea when it was developed in the late 50s. Nowadays, more and more countries, including France, prefer the notion of "reversible" deep geological disposal. In such a facility, everything is designed so that the packages can be disposed of permanently, but they can still be retrieved over a given period of time, say two or three hundred years. It is designed as a final disposal solution, the best available given the current state of our knowledge, but without precluding the possibility that our successors will find an even better solution in the future.

For practical and economical reasons, HLW packages must be surface stored for a few decades, to allow for the decay of their activity and their surface temperature, before being hauled down into a deep geological disposal facility.. They can then be stacked together more closely to save on space, especially as underground storage does not come cheap!

In Sweden and Finland, the decision was taken to dispose of unreprocessed spent fuel assemblies directly in deep geological disposal sites (in granite). Spent fuel is first encapsulated in these massive copper containers.

Most HLW activity comes from the fission products and a later stage from long-lived actinides (uranium, neptunium, americium and curium).

When spent fuel is reprocessed, the recovered uranium and plutonium remain in the nuclear cycle. Residual U and Pu traces together with the fission products and "minor" actinides are vitrified to form glass blocks, the HLW packages. As they contain only traces of uranium and plutonium, HLW glass packages cool down much faster than spent fuel. If the spent fuel reprocessing is taken a stage further to recover the minor actinides, the so-called partitioning process, then curium can be left to decay by itself while neptunium and americium can be recycled in the core of suitable nuclear reactors where they will be "transmuted" into fission products.

Vitrified HLW packages with very small quantities or traces of actinides would cool down even faster, which would reduce their containment time. Partitioning and transmutation have only been demonstrated at laboratory scale and they are still far from being ready for industrial application.

The use of such techniques would not remove the need for geological disposal, but it would reduce the design constraints that disposal facilities are subject to (see illustration page 147)

ECP 202 TW
ECP 209 TW
ECP 204 TW
ECP 207 TW
ECP 205
D
F

Future Nuclear Systems

Phenix reactor

Why Change Something Which Works Well?	150
Generation IV	150
Nuclear Reactor "Generations"	153
VHTR, Very High Temperature Reactors	151
FBR, Fast Neutron Breeders	152
Superphoenix, the Sad Demise of a Champion	153
The Six Generation IV Concepts	156

As they used to say in the Old West, "If it ain't broke, don't fix it!"

Why Change Something Which Works Well?

One of the sayings that has come down from the Old West is "if it ain't broke, don't fix it!" Light water reactors, described in chapter 5, which make up most of the reactors currently either operating or under construction, work well. They generate baseload electricity at costs which are competitive and stable, and not affected by the ups and downs of hydrocarbon prices. They are reliable and have proved to be more flexible than originally thought when they were designed. They are also very safe when run by well trained, skilled operators. The Three Mile Island accident (chapter 11) also showed that they were well capable of withstanding extreme conditions, even as far as a partial core meltdown.

They do, however, have a few limitations, which might not be too important for the moment, but which are liable to pose more of a problem as nuclear power develops in the world:

- They are only competitive in large unit sizes
- They are safe, but need highly complex systems of controls and safeguards, which require highly-skilled operators
- They were designed solely for generating electricity, while electricity only provides a sixth of the world needs for final energy
- They use less than 1% of the potential energy content of the uranium mined to fuel them
- The radioactive waste they produce includes long-lived elements (mostly actinides) which make waste disposal that much more complicated.

As long as nuclear power is only used in industrialised countries and in a few large emerging countries to power major consumption centers and/or large interconnected transportation networks, then these limitations are not significant. However, with the increasing need for sustainable development (see chapter 14), nuclear power is set to play a much bigger role, and facilities will be built in a variety of countries and for uses other than electricity generation. We therefore now need to start designing nuclear systems that do not have to operate under these restrictions. These are the generation IV systems, where "system" is taken to mean both the nuclear reactor and its fuel cycle.

Generation IV

During the 60s and 70s nuclear power experienced remarkable growth worldwide. By contrast, between 1986 and 1998, expansion ground to a virtual halt, except in Asia (see chapter 7). Recently, however, there has been a change of tack and we are seemingly on the verge of a new period of expansion, with international initiatives to prepare the nuclear systems for 2040 and beyond. In 1999, for example, the United States launched the generation IV Initiative, and was joined by nine other countries potentially able to supply the technology, along with the European Union. A year later, following a request by Russian President Putin, the IAEA, the International Atomic Energy Agency, launched the INPRO Program, which is more focussed on setting out potential user needs for 2050.

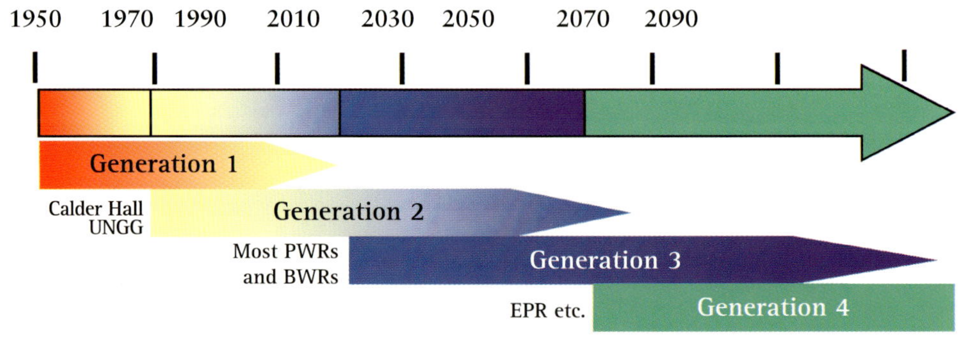

The "Generations" of nuclear reactors

The first generation covers the prototypes and early models that began operation in the 50s and 60s. The second generation is made up of the plants which currently generate 16% of the world's electricity...

The Different Generations of Nuclear Reactors

The first generation covers the prototypes and early models that began operation in the 50s and 60s. The second generation is made up of the plants which currently generate 16% of the world's electricity. Generation III, of which EPR is an archetype, has just reached the construction stage. Designed along the same technological lines as today's PWRs and BWRs, generation III reactors run even less risk of severe accident and are built to vastly reduce any external consequences should such an accident occur. The objective for Generation IV reactors is to make nuclear power sustainable by 2040 and beyond, by overcoming the limitations listed above. Specifications for Generation IV are as follows:

- Continue to safely produce inexpensive electricity, but also provide for other applications, like seawater and brine desalination and the production of hydrogen as a feedstock for synthetic fuels.
- Allow as many countries as possible to benefit from the technology, through a safety system of reduced complexity, with reactors of smaller unit size but which remain competitive and which also offer effective protection against the proliferation of nuclear weapons and terrorist aggressions.
- Make full use of the huge amount of energy contained in the fissile and fertile materials, uranium and thorium.
- Produce less long-lived radioactive waste, by keeping actinides within the fuel cycle.

The Six Generation IV Concepts

After reviewing over a hundred systems, the Generation IV International Forum, GIF, selected six concepts, described on pages 156-157, which will help focus international R&D on developing these systems for commercial application by 2040.

Although the selection of the six systems was not solely on technical grounds, two strong trends emerged:

The Very High Temperature VHTR reactor addresses the first two specifications given above, as it is designed with the high core outlet temperature of its coolant (making it suitable for hydrogen production), while the all-refractory core makes it virtually proof against melt down.

The three Fast Neutron concepts and the Molten-Salt reactor primarily meet specifications 3 and 4, fissile material economy and long-lived waste reduction, as the specific neutron-physics features of the cores mean they can answer both requirements at the same time. The Gas-cooled Fast Reactor is perhaps the most attractive long-term prospect but its fuel has not yet been developed, while the Molten Salt Reactor design is even more futuristic.

We will be looking in more detail at the Very High Temperature and the Sodium-cooled Fast neutron systems.

*Those six concepts are:
3 fast neutron breeders,
1 very high temperature reactor,
1 reactor cooled by supercritical water and 1 liquid fuel (molten salt) reactor.*

VHTR, Very High Temperature Reactors

These are the descendants of the HTR reactors described in chapter 5, but with the "V" added. HTR design was based on the "coated particle", a unique fuel developed in the UK. It comes in the form of small coated particles (with its coating, a particle measures only 1 millimeter in diameter), making it easy to adjust the quantities dispersed within the moderator, which comprises a graphite mass. The resulting core is highly refractory and is cooled by a flow of helium at high temperature, which makes it possible to produce thermodynamic cycles of very high efficiency, like those in the best combined-cycle gas turbines. The particles can be mixed in proportions that can accommodate almost every possible fuel cycle.

The HTR equivalent of a fuel assembly comes in two varieties: a prismatic graphite block in which gas channels are bored and a spherical graphite pebble.

A pebble-bed HTR, seen from above, showing the pebbles and the control rods.

Micrograph of a coated particle

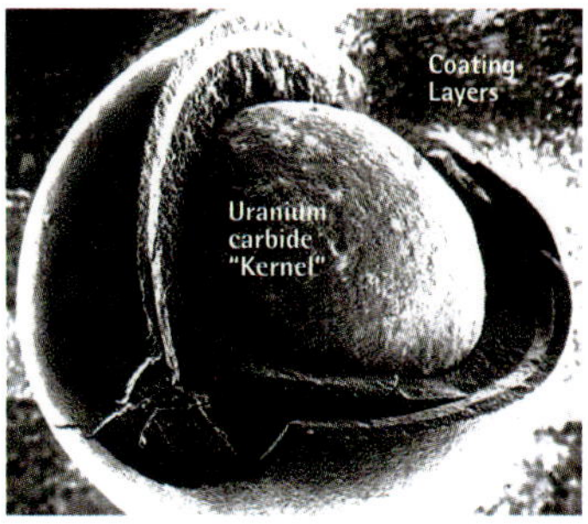

Pebbles and Prisms

As HTRs have high thermal inertia and operate well within the technological limits of their core, they are extremely safe.

In the past, there was little demand for uranium: Its only use was as a colorant for Bohemian glassware (19th century pharmacy flasks).

During the NERVA nuclear rocket propulsion project, an HTR was tested which ejected hydrogen from the reactor at over 2500°C (chapter 9).

On the downside, however, HTRs have a low power density, which leads to high capital costs. After some promising demonstrations, HTRs got off to a false start in the md-70s in the USA and in the late 80s in Germany. At the time, they produced electricity only, with a water/vapour secondary cycle which did not take full advantage of their high coolant temperature. They are currently being re-designed as small modular reactors using direct-cycle helium turbines. Their size has been computed to allow for direct passive removal of their decay heat. This is the path followed, for example, by the South African PBMR Pebble Bed Modular Reactor project.

VHTR should enable the helium outlet temperature to be pushed over 1000°C, a temperature suitable for hydrogen production

VHTR should increase the helium outlet temperature beyond 1,000°C, mainly to allow hydrogen production either through high temperature electrolysis or through catalytic thermolysis of the water molecule. A number of thermal-chemical cycles are being studied with this in view.

Whether hydrogen is considered as an energy vector, complementary to the electricity vector but easier to store, or as a feedstock for synthetic fuels to replace oil products, it seems set to be one of the solutions for the future (chapter 14). However, if we want hydrogen to be a major player, we have to find a way of producing it without giving off CO_2, in other words find an alternative to the current methods involving methane reforming or oxidation of oil products. VHTR could be the ideal solution (no CO_2 is emitted), producing hydrogen and electricity at the same time.

Breeding is based on the neutron balance shown in the table below:

	LWR	SFR
Number of neutrons released by 100 fissions	250	290
Neutrons needed to induce the next 100 fissions	100	100
Neutrons lost by leakage or sterile absorptions	90	60
Neutrons available to convert 238U into plutonium	*60*	*130*

SFR, Sodium-cooled Fast Neutron breeder Reactor

Fast neutron reactors operating as breeders can use successive recycles to exploit virtually all the potential energy content of uranium (about 100 times more than with an LWR). Better still, they can reuse the depleted uranium left over from the isotopic enrichment process. It can be seen from the table below that each time an LWR burns 100 fissile atoms, it can at best produce 60 new ones from its fertile feedstock. In a fast neutron breeder core, it is possible to create more fissile atoms than are destroyed: what actually happens is that the plutonium breeds in the core while the uranium 238 is burnt. If the core load is made to have a low breeding ratio (or even zero breeding for the "iso-generator"), where the number of fissile atoms bred is the same as the number destroyed or slightly higher, then the left-over neutrons can be used to transmute the actinides and thereby significantly reduce the long term radio-toxicity of the radioactive waste.

Liquid water moderates neutrons and so it cannot be used to cool a breeder core. The choice lies between gases and liquid metals. The only fast breeders for which we have a significant return of experience are, or were, cooled by molten sodium, at between 300 and 500°C. Although sodium is an excellent coolant, which will not corrode stainless steels in the absence of oxygen, it burns in air and reacts violently with water. To prevent any possible reaction between radioactive sodium from the primary circuit and water from the turbine circuit, current SFRs now feature an additional non-active intermediate sodium circuit. The sodium circuits are expensive, due to the materials they are made of, and this makes the capital cost of SFRs significantly higher than for a PWR of the same output. SFRs can therefore only be considered attractive when their main advantage, that of using all the energy content of uranium, is of major importance, and this of course depends both on the direction the development of nuclear power will take in the future and on the amount of uranium remaining in reserve.

SFRs met with a fair amount of success at the beginning of the nuclear age, when uranium was thought to be rarer than it actually is (at the beginning its only use was for coloring Bohemian glassware and so there was not much call for uranium exploration). The same situation cropped up again in the mid-70s when, after the first oil crisis, it was predicted that the world nuclear reactor fleet would reach 2000 GWe by the year 2000. The current prospect of rapid new growth in nuclear power explains why fast neutron breeders figure prominently among generation IV concepts.

SFRs have provided a focal-point for anti-nuclear protest in just about every country where they have been built. The main arguments against the reactors stem from the prospect of sustainable (although not, strictly speaking, renewable) nuclear power, without having to worry about the availability of mineral resources and, most importantly, the fact that people are often genuinely afraid of plutonium, which is mistakenly thought to be "the most toxic substance on Earth", and also of the increased risk of nuclear weapon proliferation (see ch.4). The fertile blanket surrounding the core of the old SFRs produced almost pure, weapons-grade plutonium 239 and it is here that the generation IV design will have to come up with a suitable solution to allay the fears of the public at large, both now and in the future.

The Demise of a Champion

In the wake of the first oil crisis, France launched a large-scale program to build 58 PWRs along with their fuel cycle facilities. At the same time, in co-operation with Italy, Germany and other European partners, France decided to build an industrial-size prototype SFR, based on the excellent return of experience from Phoenix, a 250 MWe experimental SFR that had been operating in Marcoule, in the Rhône region of France since 1973. At the time, nuclear programmes announced worldwide would have amounted to 2000 GWe by the year 2000 and a nuclear fleet of that size would have consumed all existing conventional uranium resources within 40 years. Under the circumstances, it seemed advisable to develop fast neutron breeders ready for production of nuclear power on a commercial scale by the year 2000, as they would solve the problem of the disruption of nuclear fuels.

Superphoenix was a prototype and made no claim to being economically competitive and it ended up way over the original budget. As a prototype, however, it fulfilled its role perfectly, providing a full-scale demonstrate of the technology and safety of the SFR system, while allowing its designers to solve the various teething problems that cropped up and which are inevitable in any new technology.

Superphoenix thereby provided an invaluable source of experience for pinpointing the improvements that needed to be implemented before building a commercial version. Unfortunately, the project was plagued by technical incidents from 1985 to 1997. Amid an increasingly hostile atmosphere, the reactor became a political football and following a maintenance shutdown, was prevented from starting up again. It also became necessary to restart the licensing procedure from scratch, changing Superphoenix's aim from "breeding" to "actinide burning".

The safety of the plant was not at issue, and most of the arguments put forward concerned the cost of Superphoenix. As it turned out, however, the shutdown proved far more costly than if the plant had continued operating until full burnup of the existing fuel. The project's foreign partners also had to be compensated by supplying them with free electricity and France was left to bear costs of decommissioning and dismantling all alone. As Superphoenix could no longer carry out planned R&D on the transmutation of long-lived radioactive waste, Phénix had to be given a thorough overhaul to allow it to take over the work.

Superphoenix was built between 1977 and 1985 at the Creys-Malville plant, on the Rhône River, and, at 1200 MWe, it still remains the largest SFR ever to have operated. Its construction met with violent opposition from European antinuclear activists, although protest by the local population was more muted.

When the Jospin government came to power in 1997, the Green Party put pressure on its coalition partners to shut down and dismantle Superphoenix, and closure was announced the same year

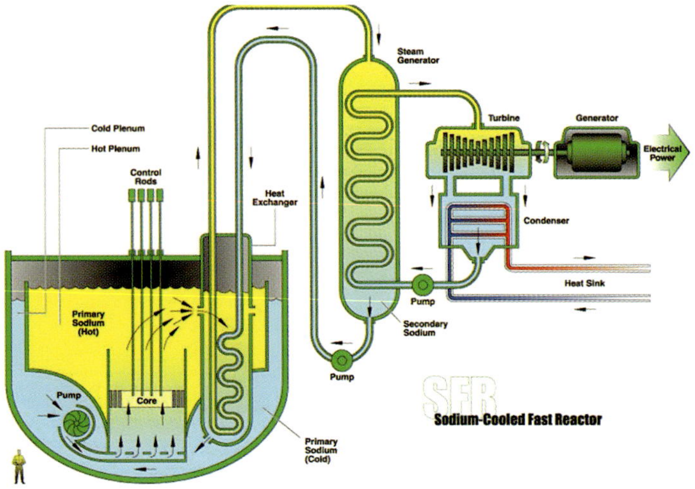

SFR, Sodium-Cooled Fast Neutron Breeder Reactor

The SFR is designed as a breeder reactor and for management of high-level waste, notably through actinide burning. With the considerable experience gained through its predecessors, the Phoenix and Superphoenix reactors in France, the SFR concept is the one that is nearest to commercial application. A demonstration plant could be built in France in 2020.

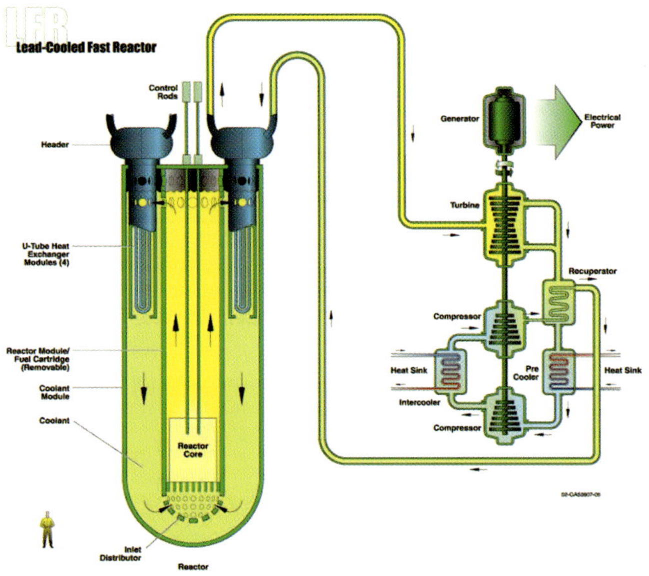

LFR, Lead (or lead alloy) cooled Fast Neutron Breeder Reactor

Plant-rating options include a small 100 MWe reactor with a pre-manufactured core which could last for 10 to 30 years and a modular reactor, rated 400 to 1200 MWe. On paper, LFRs offer the same advantages as SFRs. Lead is less chemically reactive than sodium but is toxic, very dense and corrosive. Only the Russians seem ready to support this concept.

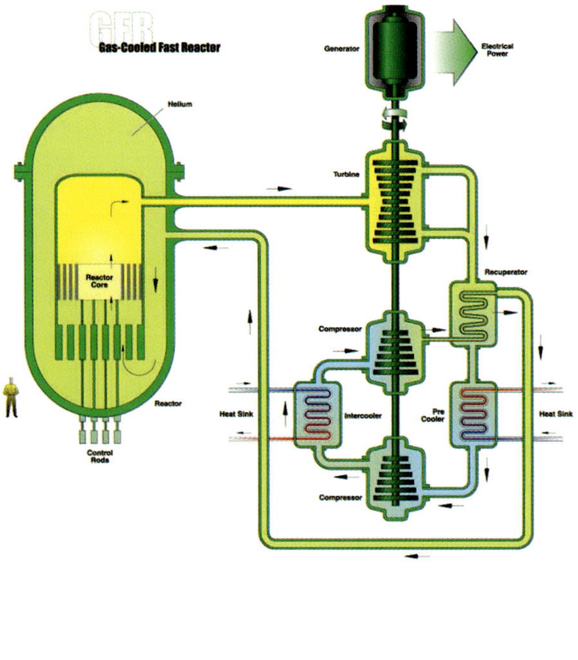

GFR, Gas-Cooled Fast Neutron Breeder Reactor

GFRs offer the same advantages as SFRs and LFRs as regards breeding and actinide transmutation and should have an excellent thermal efficiency as they are cooled by high temperature helium. Although the concept is promising, it is still in the design stage and its fuel remains to be developed.

MSR, Molten-Salt Reactor

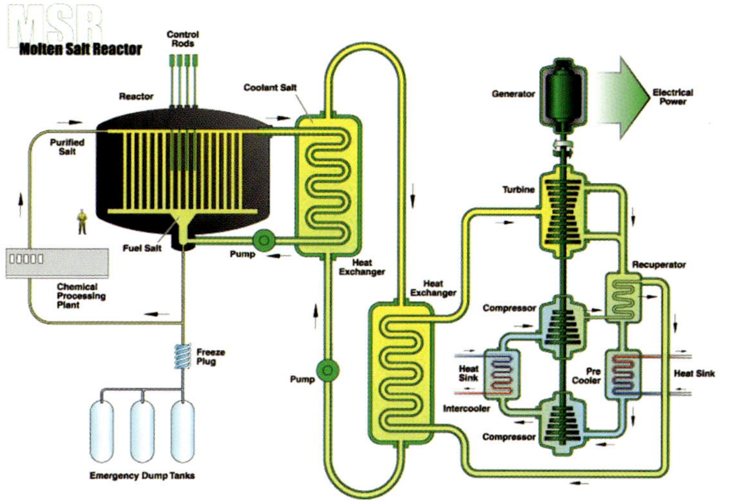

The most futuristic concept of them all, the SFR uses its liquid fuel as a coolant flowing in the primary circuit. When the molten salt flows through channels across the graphite core, it becomes critical and produces energy. The salt needs to be continuously cleaned and a reprocessing plant has to be built alongside the reactor to purify the fuel. The MSR would be best suited for a thorium cycle. The main problem is to find materials to withstand the temperature range the MSR works at.

SCWR, Super-Critical Water-Cooled Reactor

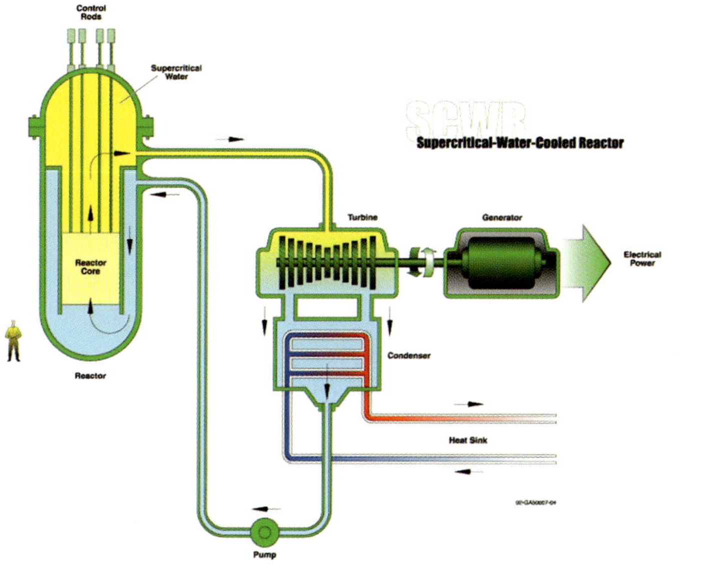

The SCWR is cooled by very high pressure water (240 bars). At this pressure, water is neither liquid nor gaseous and has special properties which make it possible to consider either a near-fast neutron breeder reactor or an open-cycle moderated neutron reactor, both with high thermal efficiency. Materials able to operate safely under those pressure and temperature conditions are still to be developed and qualified..

VHTR, Very-High Temperature Reactor

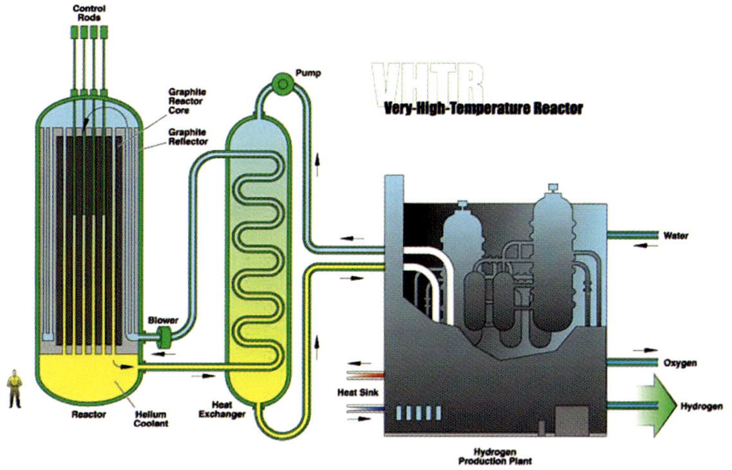

The VHTR concept uses a graphite moderated reactor cooled by helium at 1000°C. Its coated-particle fuel was developed in the 70s for the HTR, while its very high temperature makes it ideally suited for co-generation of both electricity and process heat. The heat can be used to produce hydrogen by catalytic thermolysis of the water molecule. With the benefit of previous HTR experience, a demonstration plant could be built in the USA in around 2015.

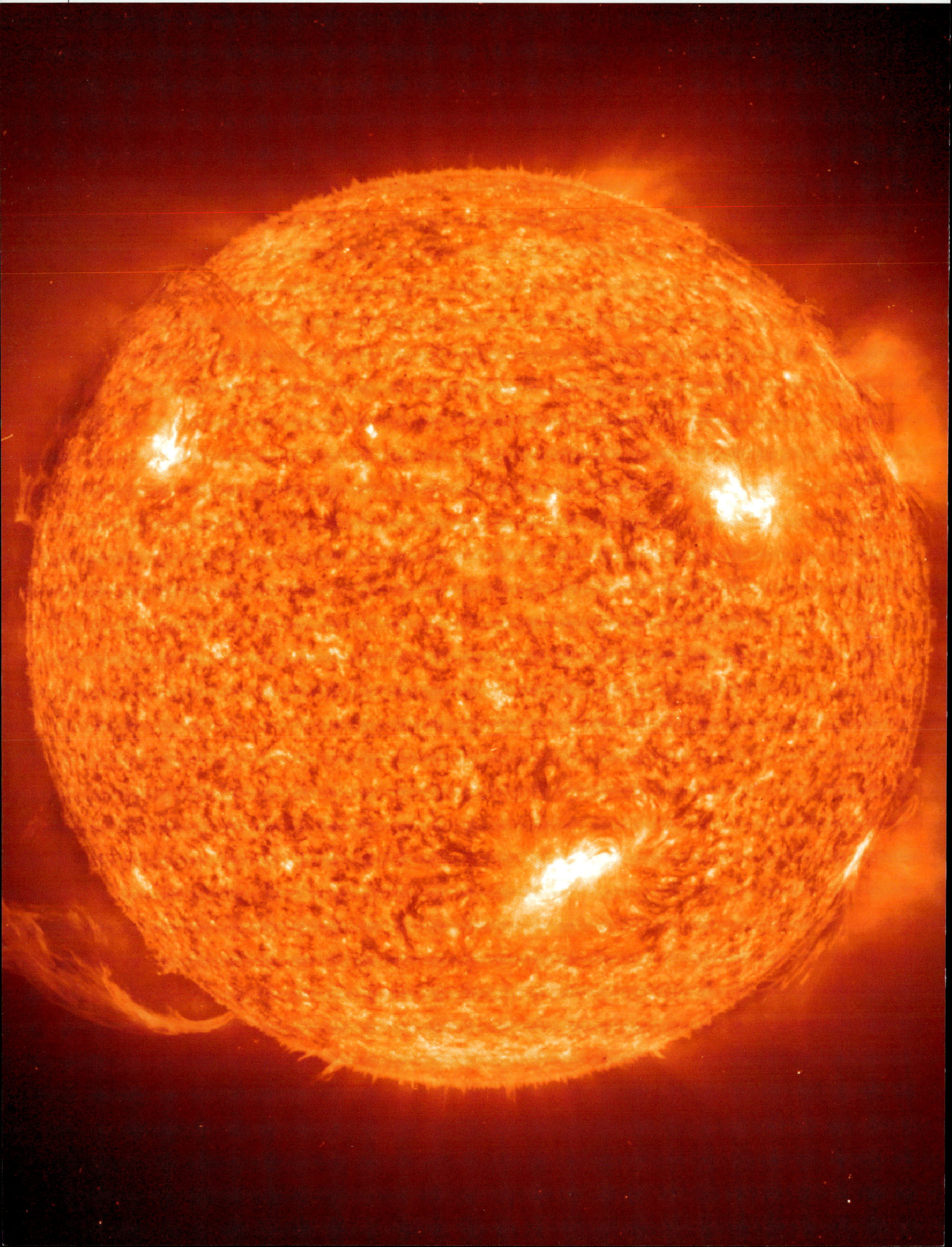

Prospects for Nuclear Fusion

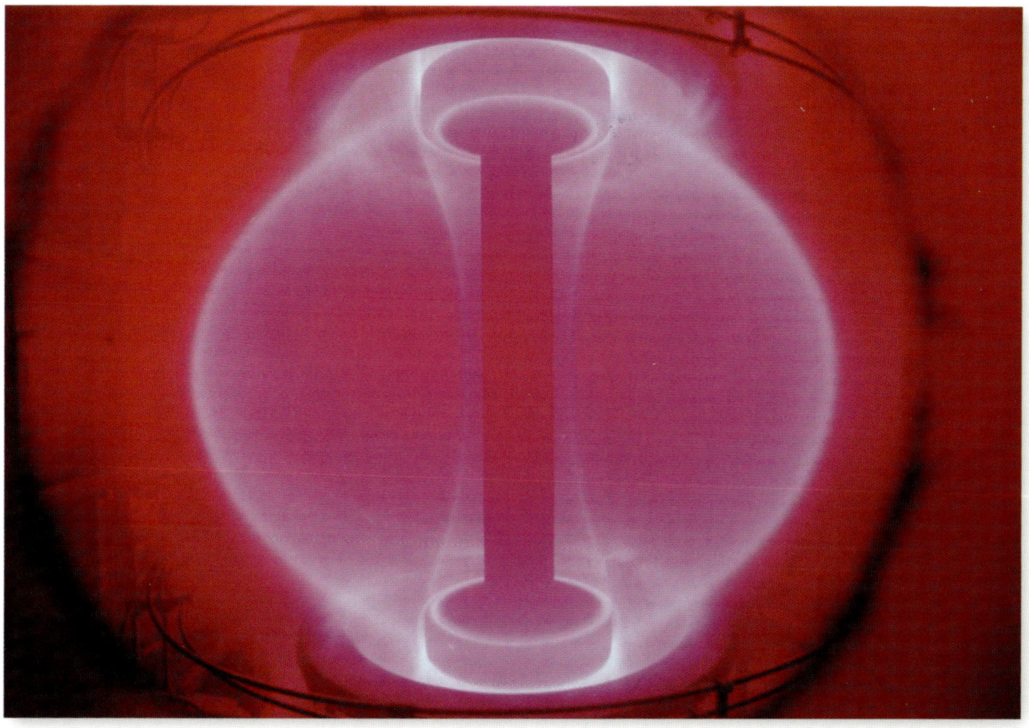

A Piece of the Sun on Earth	156
The Origin of Nuclear Energies	156
Thermonuclear Weapons, or H-Bombs	157
Nuclear Fusion: Theoretical and Practical Aspects	158
Cold Fusion	158
Fusion in the Sun and on Earth	159
Magnetic Containment Fusion (MCF)	160
The ITER Project	161
Inertial Containment Fusion (ICF)	162
Prospects	164

A Piece of the Sun on Earth?

As can be seen in the inset below, there are two large sources of nuclear energy in the Universe, both resulting from a loss of mass:

- Radioactivity energy, the source of the Earth's internal heat,
- Fusion energy, the source of the heat radiated by the Sun and other stars.

The Origin of Nuclear Energies

Everybody knows Einstein's famous formula $E = mc^2$, which should, strictly speaking, be written:

$$\Delta E = - c^2.\Delta m$$

This formula shows that when a body has mass, it has a huge amount of energy associated with this mass and that a small variation in it corresponds to a huge energy variation.
Binding energy is considerable and, as we already saw, is about a million times higher than the energy involved in chemical reactions between atoms and molecules.

As explained in chapter 1, almost all the energy sources available on Earth are either directly or indirectly of nuclear origin : solar power, wind power, hydropower (partly), geothermal energy, biomass and all the fossil fuels originating from biomass (coal, oil and natural gas).

The First Fusion Reactions

It could be said that Man's control of fusion reactions comes down to creating "a piece of the Sun", on Earth as the reactions are similar to those which produce Sun's enormous power.

However, when Man first managed to produce fusion reactions, in 1952, it was not through a controlled reaction, but by exploding a thermonuclear weapon, or "H-Bomb" (see inset opposite). Ever since then, vast teams of researchers have been involved in a concerted worldwide effort (rare enough in itself) to "tame" the reactions.

A small fraction of the energy sources available on Earth is not of nuclear but of gravitational origin : tidal power and part of hydropower

Thermonuclear Weapons, or "H-Bombs"

After the Trinity test saw the first (fission) atomic bomb explode in the New Mexico desert in July 1945, the USA was for a number of years the only country to possess nuclear weapons. Helped by the atomic spies Fuchs, Pontecorvo and others, it took the Soviet Union only four years to develop their own weapon and carry out their first atomic test, in 1949. During the ensuing crisis, which ushered in the McCarthy era and the communist witch-hunt, hawks dominated the US State Department and the Pentagon and cleared the way for the development of a Superbomb, better known as the hydrogen or H bomb, in which a conventional atomic explosion is used to trigger a thermonuclear fusion bomb.

The secret of the H Bomb, which is hundreds of times more powerful than the fission bomb, was discovered by Ulam and Teller, and the Americans made enormous efforts to stop the Soviets getting hold of its design. All their efforts proved to be in vain, however, and the arms race again became a two-country affair when the USSR tested its own H bomb just 12 months after the 1952 US test.

Under Kruschchev, the Soviets carried out the most powerful atomic explosion ever seen, estimated at 60 megatons, - the equivalent of 60 million tonnes of chemical explosive TNT or 3,000 times more powerful than the Hiroshima bomb.

Under Kruschchev, the Soviets carried out the most powerful atomic explosion ever seen, estimated at 60 megatons, - the equivalent of 60 million tonnes of chemical explosive TNT or 3,000 times more powerful than the Hiroshima bomb

Uncontrolled fusion
H-Bomb

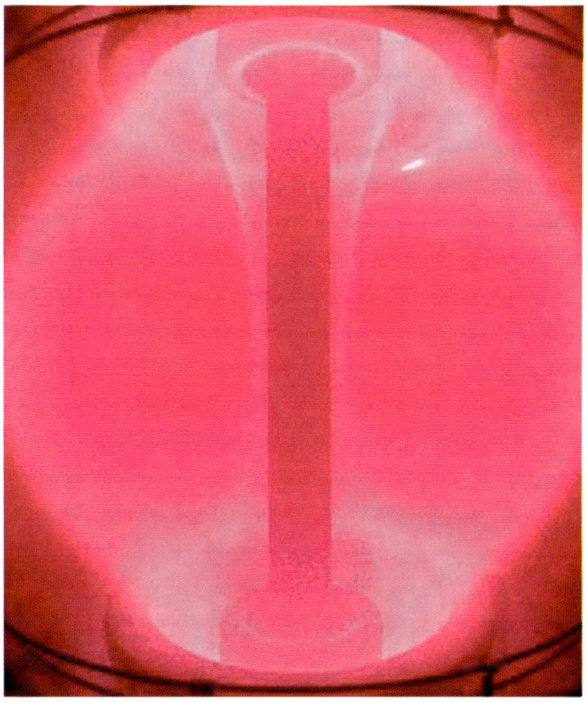

Controlled fusion
Tokamak facility

Although nuclear binding forces are enormous, they only act at very short range

Nuclear Fusion: Theoretical and Practical Aspects

The most stable combination of a group of nucleons is a medium-sized nucleus. This means that if a heavy nucleus can be split into two medium-sized nuclei, binding energy will be released. This is nuclear fission, the main subject of this book. Similarly, if two light nuclei can be fused to form a medium-sized nucleus, the binding energy can also be harnessed and this is what we call nuclear fusion.

Although nuclear binding forces are enormous, they only act at very short range. This means that to achieve fusion, two positively charged nuclei have to be brought together close enough for these forces to exceed the electromagnetic force keeping them apart. This is called "overcoming the Coulomb barrier".

In the Sun and other stars, it is the gravitational force that drives nuclei together, overcoming the Coulomb barrier, and fusing hydrogen nuclei into helium nuclei, helium nuclei into carbon nuclei, and so on. Fusion is at the origin of solar energy.

Relatively speaking, the "easiest" fusion to achieve is between deuterium and tritium, both isotopes of hydrogen. Deuterium (D) has a nucleus which is made up of one proton and one neutron, and it exists on Earth in tiny proportions in natural hydrogen. The nucleus of tritium (T) is made up of one proton and two neutrons. Tritium does not exist naturally as it is radioactive, and has a short decay half-life (around 12 years). and therefore has to be produced by bombarding an isotope of Lithium with neutrons. D-T fusion produces a helium nucleus (α particle) and a neutron, and results in a considerable release of energy.

Controlling fusion here on Earth, without the enormous gravitational pull of the Sun, for example, to help confine the plasma of charged nuclei, is an extremely difficult undertaking.. Even nowadays the only way to achieve large scale fusion is to transfer the huge energy of an atomic explosion to a mixture of deuterium and tritium. This is the principle of the H-Bomb developed in the early 50s (see previous page). Achieving controlled fusion is much more difficult. Scientists all over the world have been working on this problem for the last forty years or so, and have been following two main lines of research: magnetic containment (MCF) and inertial containment (ICF.

About 15 years ago, it was thought that fusion might be achieved through somewhat less complicated processes (see inset below), but this proved to be a false hope.

Cold Fusion

The term "cold fusion" was minted to describe a process which could control nuclear fusion reactions at room temperature. In March 1989, two scientists, Stanley Pons and Martin Fleishman, reported at a press conference that they had achieved a cold fusion process with a net release of energy, by electrolysing heavy water (deuterium oxide) with specially conditioned palladium electrodes. This announcement raised quite a ruckus among nuclear physicists and many teams around the world tried to reproduce Pons and Fleishman's experiment, but to no avail. Once the initial dust settled, a few teams went on trying to achieve cold fusion, through electrolysis or gaseous discharges, and some of them did experience energy release, but on a very small scale and the process was not very reproducible. Any hope that a useful source of energy had been discovered has now been abandoned and cold fusion is no longer a topic of conversation. However, no one has as yet satisfactorily answered the question of how the energy release came about.

Nuclear Fusion

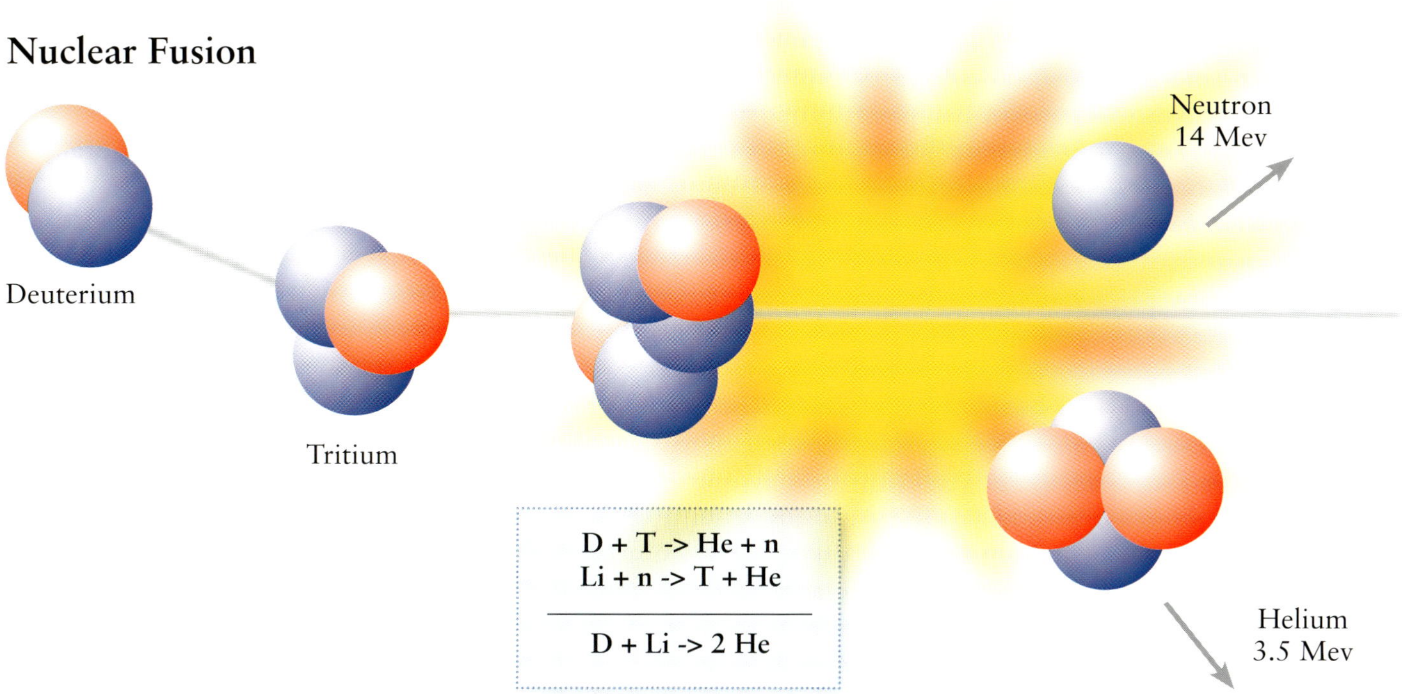

Fusion, in the Sun and on Earth

	Gravitational Containment, the Sun	Magnetic Containment, ITER	Inertial Containment, Laser MJ
Duration	10 billion years	15 minutes	0.05 billionths of a sec.
Density	~1.4 g/cm³	0.005 billionth g/cm³	300 g/cm³
Temperature	5,800 °C surface 15 million °C center	100 million °C	100 million °C
Power Density	0.01 w/m³	500.000 w/m³	**
Characteristic distance*	1 million km	10 m	1 cm

* diameter of the "reaction zone" ** has no meaning over such short duration

Solar Energy

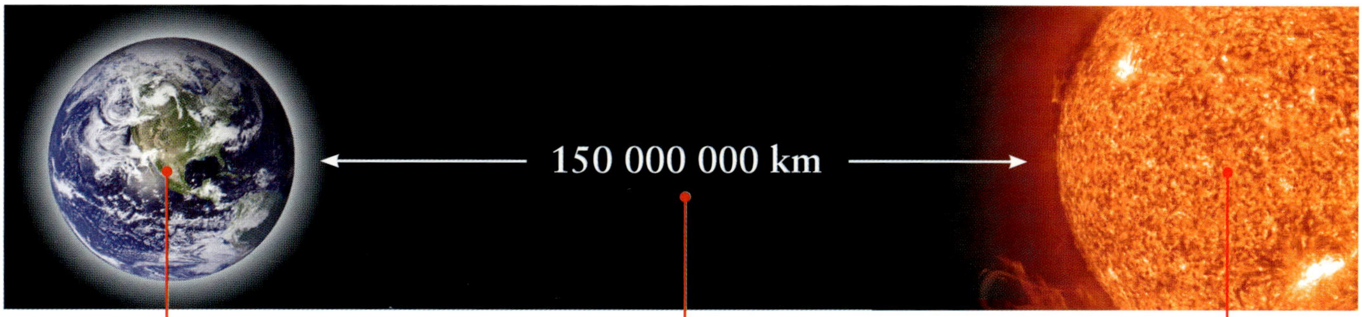

The inner temperature of Earth, where geothermal energy comes from, is due to the decay of radioactive elements that have been there since Earth was formed.

Earth's surface receives 10,000 times more energy (nuclear fusion energy) from the Sun than it receives from its core (energy of natural radioactivity).

The heat radiated by the Sun is produced by nuclear fusion reactions.

Magnetic Containment Fusion

Since they were first invented, Tokomak design has come along in leaps and bounds and they can now achieve huge temperatures

The first method for forcing deuterium and tritium and nuclei together and fusing them involves using thermal agitation to accelerate the nuclei to enormous speeds. This means heating the plasma to 100 or 150 *million* degrees! By way of comparison, the Sun's surface has a temperature of only 5,800°C.

This ultra-hot plasma cannot of course be allowed to come into contact with any kind of material wall. The solution is to contain the plasma within a sort of magnetic bottle through a combination of an electric field and extremely powerful magnets. The most widely used machine for doing this today is the Tokomak, designed in the Soviet Union in 1954 by a team led by Lev Artsimovich. The most powerful Tokomaks operating today are the European JET, located near Oxford, UK, and the Japanese JT 60. Inside a Tokomak, very low density deuterium-tritium plasma (100,000 times less dense than air at atmospheric pressure) is contained within a toroidal-shaped chamber by a series of magnetic coils and heated by the Joule effect, as it conducts electricity.

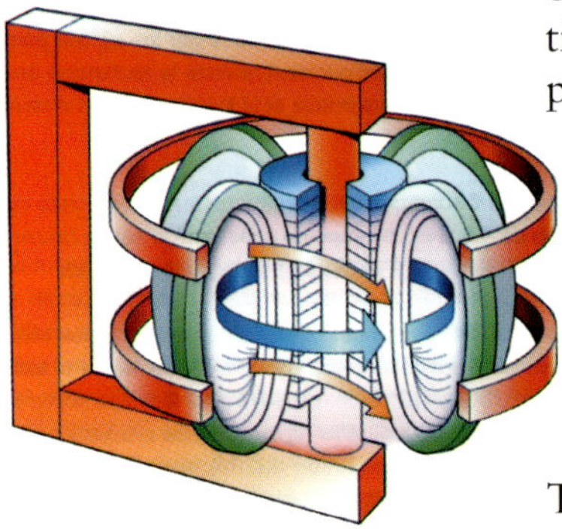

Since they were first invented, Tokomak facilities have experienced progress in perfomances ranging over several orders of magnitude. However, each time the scientists manage to get over one obstacle, another rears its head and there are still a lot of difficulties to overcome. Other magnetic containment devices are under development, notably the Stellarator, which can operate continuously rather than in pulses, but the Tokomak is the most advanced system currently available.

The latest Tokomak plasma fusion device is the ITER Project, which will be twice the size of JET. It is to be built by a 7-country consortium, a project which can really only be compared in scale to the International Space Station.

View of the inside of JET's plasma chamber (before commissioning)

The ITER Project

On June 28, 2005, seven countries (European Union, Japan, Russian Federation, United States, China, South Korea and India) decided to get together and build the world's largest Tokomak, in Cadarache (France). ITER, the project's name, comes from the Latin word for "the way", and actually stands for the International Tokomak Experimental Reactor. The facility should begin its experiments by 2016 and remain operational for at least 15 years. The objective is to demonstrate the scientific and technological feasibility of a full-scale fusion power reactor by momentarily producing ten times more thermal energy from fusion heating than is supplied by auxiliary heating.

It took twenty years between the international agreement giving the go-ahead for ITER signed at the 1985 Reagan-Gorbachev-Tanaka-Mitterrand Summit and the eventual decision to locate ITER at Cadarache. It will need another ten years before the first plasma. is produced. When this happens, this plasma will be about the equivalent for fusion of what the first chain reaction in CP1 on December 2, 1942 was for fission.

This first ITER plasma will be about the equivalent for fusion of what the first chain reaction in CP1 on December 2, 1942 was for fission

Before the first demonstration reactor is built, however, it will need more than just a successful ITER experimental project. There needs to be a complete program, alongside the project, to develop material for the confinement vessel which will be able to stand up to the very high temperatures the plasma needs, as well as the constant bombardment by very high energy neutrons.

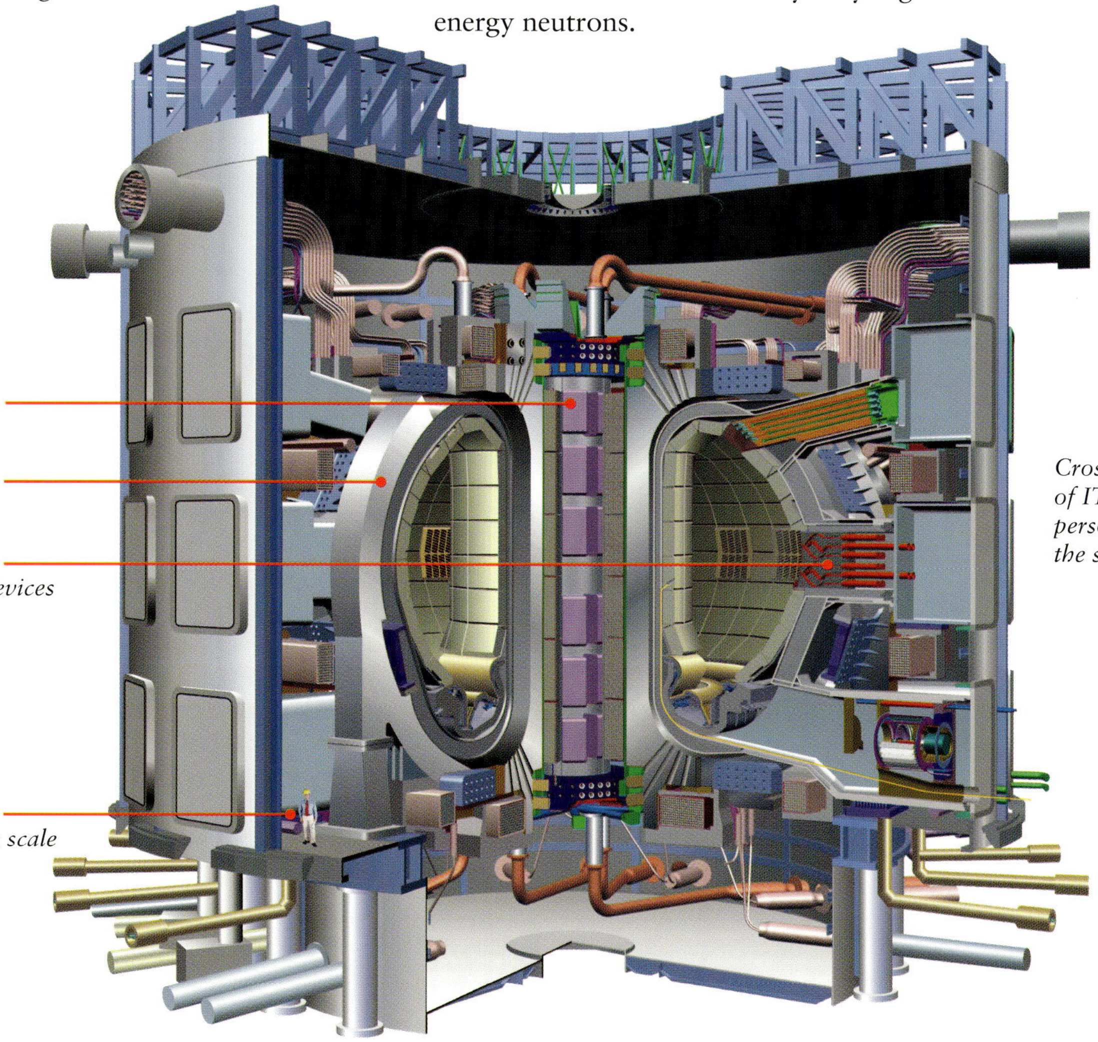

Cross-section of ITER (the small person in blue shows the scale).

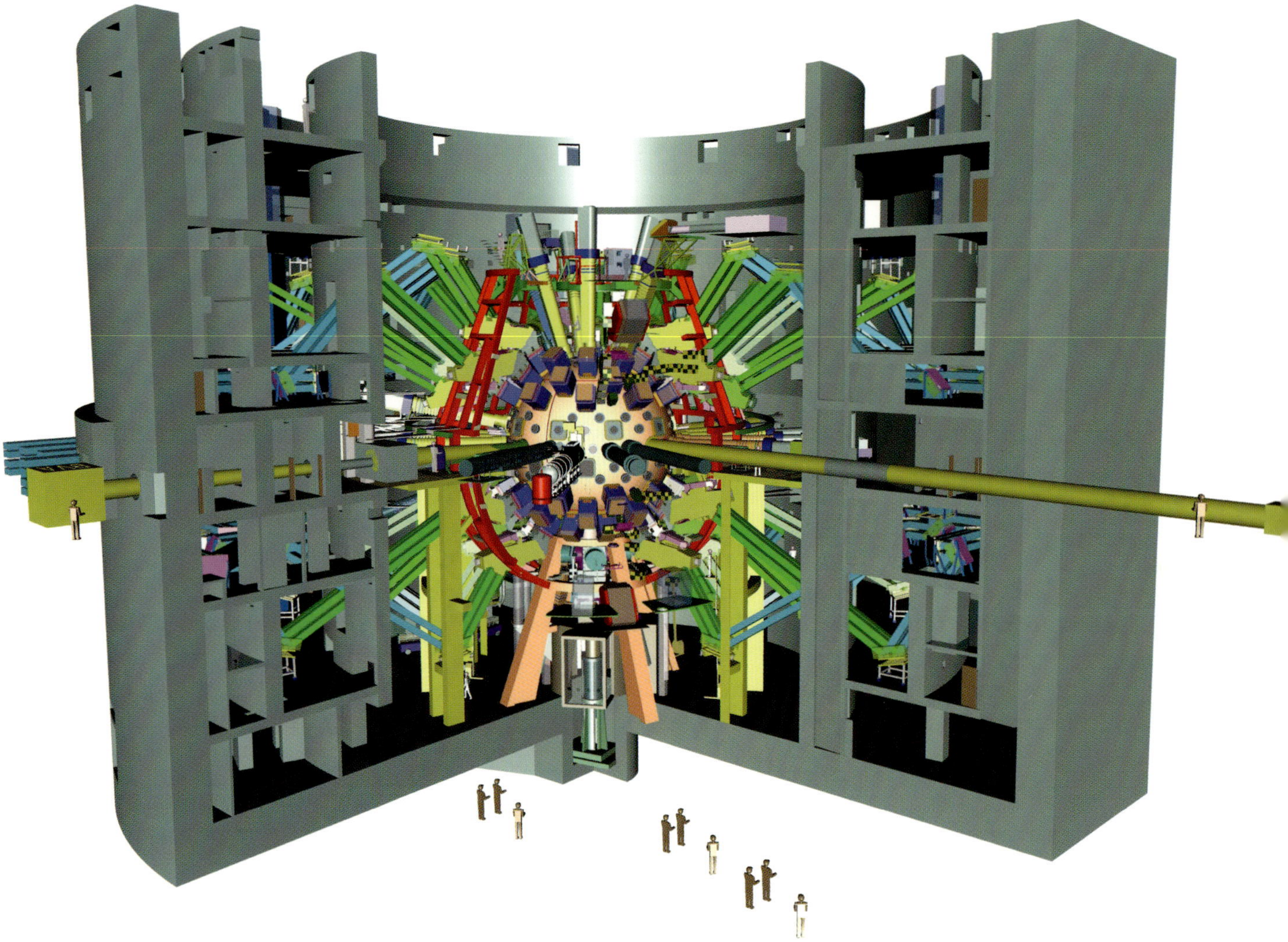

Cutaway of the reaction chamber building of "Laser Mégajoule" (240 laser beams)

Inertial Containment Fusion-ICF

Another technology is being developed to achieve controlled release of fusion energy. This approach involves compressing a tiny pellet filled with a deuterium - tritium mixture to very high densities (1000 times the density of a solid). Energy is delivered to the target using about 200 laser beams of very high intensity (direct drive). The external surface of the pellet is instantly volatilized, causing a violent shock wave to compress and heat the mixture inside to a temperature where fusion can occur.

The laser beams can also irradiate a gold cylinder that surrounds the target (indirect drive). Under laser irradiation, the internal surface of the cavity shoots very hard X rays at the target to volatilize its surface. This process is somewhat similar to that of an H-bomb, but on a much-reduced scale.

Research on ICF has been mostly carried out for military purposes, especially after the Comprehensive Test Ban Treaty forbidding nuclear weapon testing. Two large experimental facilities, of similar design, are presently under construction. One, the NIF (National Ignition Facility) is being built near Livermore in California and the other, the Laser Mégajoule, near Bordeaux in France.

We are still very far from being able to build an ICF power reactor. To generate 1 GWe we would need a laser able to implode 10 targets per second, with 5 megajoule impulses each.

(1) NIF : National Ignition Facility

The National Ignition Facility (NIF) at Livermore (USA): Installation of the reaction chamber.

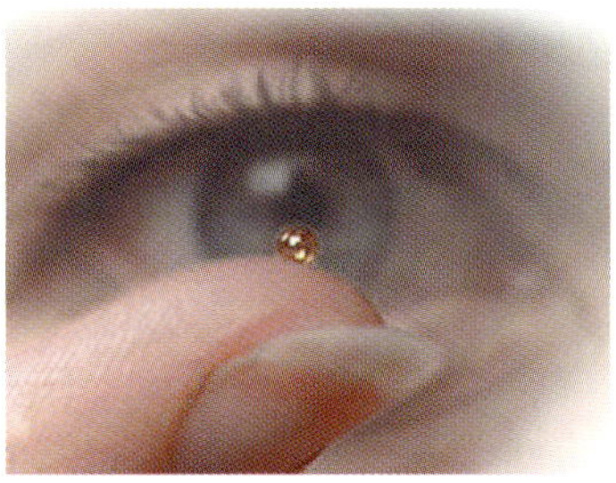

Nova target (NIF targets will be roughly double in size).

The target-holder of NIF, seen through one of the openings through which one of 192 laser beams will enter the reaction chamber to focus on the target.

Some people believe that fusion will replace fission, but this is far from certain

Prospects for Nuclear Fusion

Fusion is still a very distant target which is very difficult to reach, but it offers very attractive prospects:

- An almost inexhaustible fuel supply (lithium resources are a potential energy source at least equivalent to coal, or the uranium used in breeder reactors)
- No risk of reaction runaway (there are only a few milligrams of plasma in the reactor core!)
- Integrated fuel cycle (tritium is produced and separated in the reactor itself)
- A significant production of radioactive waste, but of short-lived waste only (from the activation of materials under neutron flux).

Q1: Are we certain that Man will eventually control nuclear fusion ?

Yes, from the experimental physics angle, the large Tokomaks currently operating have begun the process, and ITER should complete the demonstration.

But the answer would be "no" as far as any practical application of the technology is concerned.

Q2: Which of MCF and ICF will come out on top?

Magnetic containment fusion currently offers more promising prospects for generating power. Inertial containment fusion is mostly used for simulating nuclear weapons. However, who knows what the future holds in store?

Q3 :Will fusion power ever be economically competitive ?

Quite frankly, nobody knows the answer today. It will certainly not be possible before the actual commissioning of a genuine demonstration reactor, around 2050. Some people believe that fusion will replace fission, but this is far from certain.

By the time fusion is able to make a significant contribution, hydrocarbon reserves will be severely depleted, and our grandchildren may find themselves happily living with two nuclear energy sources, the generation V or VI fission reactors and fusion reactors.

The discovery of oil did not do away with the need for coal, and it is likely that fission and fusion will, in the long term, turn out to be complementary (in terms of uses, ratings, etc.) rather than competitors.

The CEA Site of Cadarache (France)

ITER, a Ride to the Future

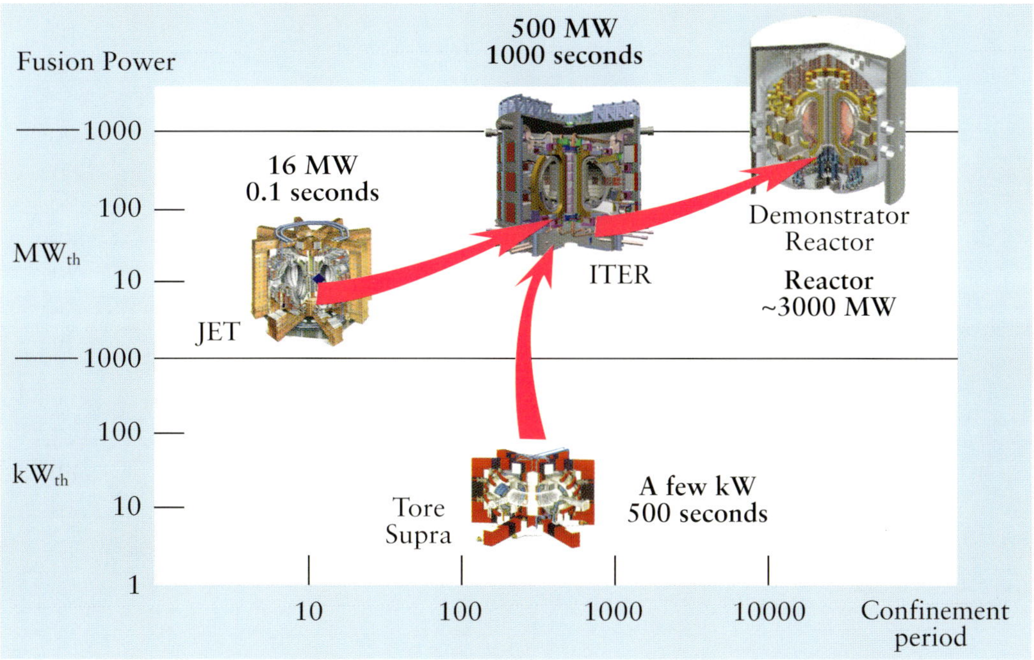

ITER will set out to demonstrate the scientific and technological feasibility of a full-scale fusion power reactor

A World Vision of Energy: 2000 - 2100

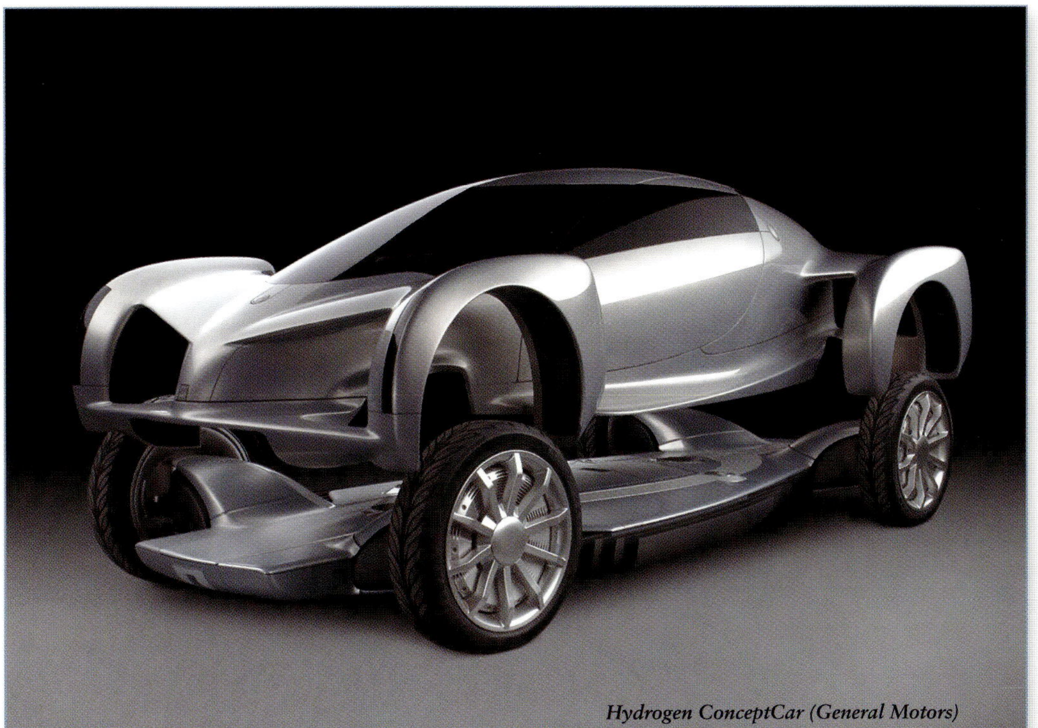

Hydrogen ConceptCar (General Motors)

Energy and Sustainable Development	172
There's More to it than Just Electricity	181
Future Applications of Nuclear Energy	182
Nuclear Energy and Water Production	183
Nuclear Energy and Heat Production	184
Hydrogen Production	185
Conclusion	186

The Earth will not be able to feed 8 to 10 billion people on a sustainable basis unless agriculture remains highly mechanized, with enough water for irrigation and spraying and enough fertilisers and other agrochemical products

Energy and Sustainable Development

There does not seem much point in discussing sustainable development in our societies if we are unable to find a lasting solution to the energy supply. Obviously this does not mean that every person on the planet needs to have the 8 toe (tons of oil equivalent of various energies) that the average American enjoys today, or even the 4 toe annual supply that the Europeans and Japanese have. People can certainly live well enough with half as much as the Europeans and a quarter as much as in the USA, say 2 toe per year per inhabitant.

However, even if we manage to get the developed countries to bring their energy consumption substantially down over the next 50 years, by the end of the 21st century we will still need some 20 billion toe of energy to meet the needs of the 10 billion or so people who will in all likelihood be living on Earth. This represents about double the energy consumption of 2005, when some 10 billion toe of commercial energy was consumed worldwide, or 11 billion toe if we include non-commercial energy sources such as wood gathered for heating and cooking, animal dung, etc.

If we want to take a realistic view of what we need to ensure sustainable development for a planet with between 8 and 10 billion inhabitants between 2050 and 2100, we need to find answers to the following questions:

- How can these populations be fed?
- How can they be provided with sufficient water?
- How can their environment be kept healthy?
- How can the climate be kept relatively stable?
- How can people have a reasonable life expectancy?

All these questions address issues that are part of the underlying causes of one of the major problems facing us in the future : the risk of massive migrations of underprivileged populations. The questions are just a small part of a wider picture which, as we will be seeing, leads directly to the supply of energy.

How are we going to feed everyone in the 21st century?

The Earth will not be able to feed 8 to 10 billion people on a sustainable basis unless agriculture remains highly mechanized, with enough water for irrigation and spraying and enough fertilisers and other agrochemical products. All of this, however, means heavy direct and indirect energy consumption for growing the food and manufacturing the products we need.

The idea of a mass return to using horses and cattle to replace tractors as a source of energy for agriculture or replacing artificial fertilizers by animal manure or vegetable compost does not stand up to a costed analysis. This type of solution might work fine for a population of 2 billion, but it would prove woefully inadequate for a population four or five times bigger. We should not forget that when the first powered machines (first steam, then oil) started to be used in agriculture, about 30 to 50% of arable land in Europe and the USA was given over to producing fodder for the horses and cattle who used to pull the ploughs and do the work around the farms. Once the oil peak has been reached, oil resources are expected to decline steadily, which means there will be considerable pressure exerted to grow crops that can be used for biofuels. This competition for the use of arable land will lend even less realism to the somewhat romantic notions of sustainable agriculture based on organic production, with its much lower yields than modern farming methods.

This does not mean that in the future we should not make more use of draft animals in agriculture for hauling machinery and for transportation. We could also make better use of fertilisers of animal or vegetable origin. The pipedream, though, is imagining that these "environmentally friendly" solutions will enable us to make huge reductions in our use of mechanized farming and nitrogen-based fertilisers. The food produced on the farms also has to be transported in pristine condition from the growing areas in the country to the towns and great cities and it would be difficult to see how we could move people out of the conurbations and closer to the food sources. The food chains are themselves heavy consumers of energy, especially the refrigeration facilities and the transport.

The dwindling energy supplies and their increasing cost will force us to put an end to energy-wasting practices such as importing out-of-season fruit from distant regions, and also oblige the richest countries to cut down on their consumption of meat and fish. We currently need about a liter of diesel fuel for each kilo of fresh fish brought into our ports, and double that for frozen fish bought in a supermarket (which is the same as about 50 kilometers of careful driving). A kilo of beef from an animal raised in a modern intensive farm needs between 3 and 4 liters of diesel to get it onto our plates (the equivalent of about 75 to 100 kilometers of driving).

When you consider our habits, our efforts at cutting down food-chain energy consumption by 50% will be way off target. Our future depends not so much on making do with less energy, but on boosting our energy sources so that we are able to feed the world's population. Farms for salmon, bass, bream and other fish are not sustainable on a large scale, especially when you consider that to produce one kilo of fish on a farm, you need 5 kilos of "smaller fish" in the form of fish meal and oils (animals grown for meat are mainly herbivores, while fish farms produce carnivorous fish).

Salmon farm in British Columbia

How can we ensure an adequate water supply?

Water is now seen as key to sustainable development. Even if people have been aware of this for quite some time, and it was the subject of considerable discussion at the Rome Club in the early 70s, it is only recently that the public at large has found out that the world faces a shortage of water in the future if we continue on the present path.

To produce one kilo of fish on a farm, you need 5 kilos of "smaller fish" in the form of fish meal and oils (animals grown for meat are mainly herbivores, while fish farms produce carnivorous fish).

A bottle of water delivered to the point of sale needs the equivalent of 100 - 150 g of oil: 30 g of plastic, which needs 50 g of oil, plus 50 to 100 g of oil for filtering, bottling and shipping the water. In a bottle, the water represents under 5% of the value of the energy.

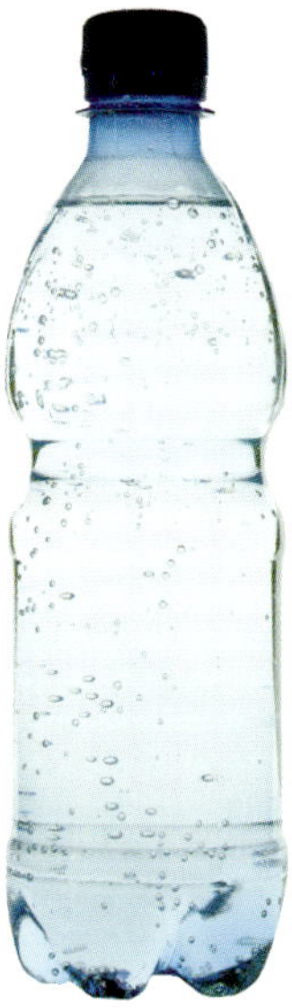

The problem of pollution and the deterioration of the quality of fresh water (rivers, lakes, aquifers, etc.) has become an issue we are all facing on a daily basis. It is no coincidence that people are turning more and more to drinking bottled water, which is no longer the preserve of the more comfortably-off in the population but has become commonplace right across the social strata.

Pollution peaks in our rivers following heavy rains or hot snaps no longer just affect anglers. The relatively common spectacle of fish floating belly-up on the surface of the water is something that is bringing the problem home to us all.

The even more important issue of overconsumption of water from reservoirs is also beginning to impinge more and more on public awareness. Examples include projects, like the great Libyan man-made rivers, that set out to use non-renewable sources of water from fossil aquifers. These projects simply pump up water that is not subsequently replenished, and will eventually empty the reserves in exactly the same way as we have seen for the oil and gas-fields.

Other examples of overconsumption include pumping out water from aquifers that are replenished by surface water (rainwater, watercourses, etc.), but in such quantities that the supplies dry up, in much the same way as fish populations decline to a critical threshold when overfishing occurs (the source is taken out faster than it can be renewed).

Can we reduce the demand for water or is it possible to increase the supply in a sustainable way? The only solutions we have come up with so far to deal with these issues are on the one hand to reduce water consumption (which penalizes agriculture as the heaviest consumer and therefore affects our food supply or the supply of biofuels) or to build desalination plants to turn seawater into fresh water.

Desalination and its possible links with nuclear energy are discussed later on in this chapter.

November 14, 2005, Dubai. A tree stands alone among the dunes.

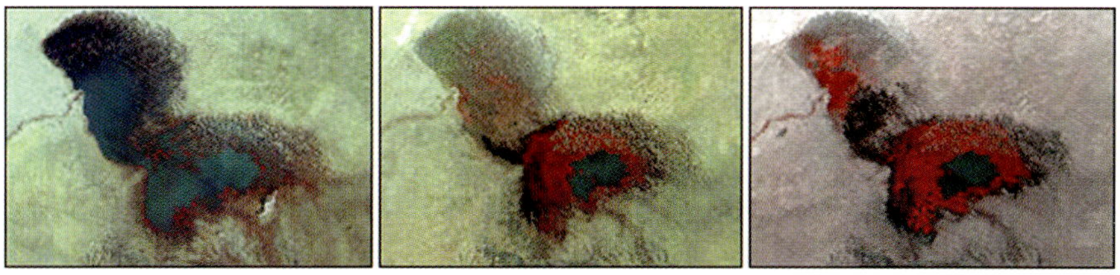

One thing people also need to be made more aware of is the fact that the water we use has an increasingly high energy content. We are having to dig further down to find it, it has to be transported over longer distances and irrigation is using more and more mechanical energy.

Water from the desalination plants or in the bottles we buy in the local supermarket all have an energy content that is some 10 to 100 times higher than the tap water we get at home.

How can we keep the environment healthy?

The industrial revolution and the demographic explosion have caused deep-seated damage to our environment. Examples include deforestation, first in the 19th century in Europe and then in Africa, Asia and Latin America, and the 20th-century plagues of accelerated soil erosion, polluted rivers and lakes, a declining terrestrial and marine biodiversity (especially alarming for the large symbolic species such as whales, rhinoceroses, tigers, bears and pandas, etc.) and marine and atmospheric pollution, which started in urban areas back in the 19th century, but which has now become a global phenomenon (acid rains, ozone hole, greenhouse effect).

A lot of this pollution stems from our consumption of commercial energies, which rose twenty-fold between 1990 and 2005. Energy consumption over this period consisted mainly of fossil fuels, which on average accounted for 75% of consumption in the 20th century and 85% now.

We have been gradually learning how we can limit or control some of the effects our energy consumption is having on the environment, but there are still large areas that remain outside our control, and a number of new problems have recently made an appearance, one of the more worrying being the change in the composition of our atmosphere as it tries to cope with the steady influx of greenhouse gases. The main effect is climate change which has now moved beyond the theoretical stage to become a reality that every one of us, and not just the scientists, can see before our very eyes. Climate change is considered to be a major challenge for the future of human societies.

What does seem evident is that, from the environmental point of view, nuclear energy offers a number of clear advantages over fossil fuels.

If every country was rich enough and had the advanced technology to develop electricity grids supplied with nuclear energy, people would no longer need to cut down trees for wood for cooking their food and heating their homes. It would also sharply reduce the need for coal for domestic and industrial use, thereby cutting atmospheric pollution resulting from ash, dust and sulphur, not forgetting the radioactivity released by non-nuclear power plants (a coal-fired plant releases

Lake Chad has been steadily evaporating and its surface area has gone down by 80% in the last 40 years. Population growth, agro-pastoral development and desertification have led to a disturbing reduction in fauna and flora

The man in the street thinks it is reasonable to be able to buy exotic or luxury products on special occasions, but it seems less reasonable for most fruit, vegetables or flowers to be flown thousands of kilometres so we can enjoy them all year round.

more radioactivity into the environment per kilowatt hour than a nuclear power plant). If we were able to come up with electric batteries with a storage capacity ten times higher than the ones currently available, we would see cars running on electricity, with highly beneficial effects on urban pollution. What also needs to be tackled here, though, is the problem of charging the batteries without producing carbon emissions.

It just needs a few simple sums to realize that renewable energies are not the catch-all solution they are sometimes imagined to be, but can only meet some 10 to 20% of our increasing energy needs (domestic and transport).

The remainder will therefore have to come from nuclear energy, which now seems set to become the driving-force behind the environmentally-friendly approach that is key to the notion of sustainable development.

How can we introduce the notion of reasonable travel

The use of the word reasonable in the title of this paragraph is sure to raise a few hackles. For the purists and philosophers among us, "reasonable" is a totally subjective notion that really has very little meaning. A "reasonable" life expectancy in the Middle Ages was 20 years, and the fact that 90% of French people in 1900 had never left their region or even their village, was also considered "reasonable". However, it could also be

Simplified diagram of commercial energies produced and consumed worldwide in 2005 and the authors' projection for 2100

2005	2100
6 Billion People	10 Billion People
World Primary Energy **10 Gtoe** Fossil energies: 86% Nuclear energy: 7% Renewable energies: 7%	World Primary Energy **23 Gtoe** Fossil energies: 40% Nuclear energy: 50% Renewable energies: 10%
Final Energy consumed **6 Gtoe**	Final Energy consumed **17 Gtoe**
Heat 3 Gtoe / Electricity 1 Gtoe / Transport 2 Gtoe	Heat 5 Gtoe / Electricity 8 Gtoe / Transport 4 Gtoe
Nuclear **440 Reactors** **350 GWe installed capacity** **800 MWe average rating**	**Nuclear** **3,000 Reactors** **6,000 GWe installed capacity** **2,000 MWe average rating**

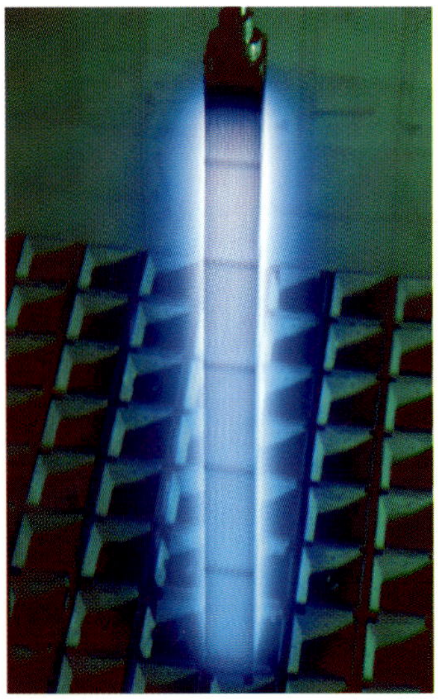

Even if we just consider local (urban pollutants) or semi-local pollution (tropospheric or low-altitude ozone), it is easy to see the role played by fossil-fuel consumption (coal or oil-fired power plants, domestic use, pollution from car exhaust emissions).

said that the average man in the street has an instinctive idea of what is reasonable and what is not.

It seems reasonable to take a long-distance flight once a year (even though such a thing is still well beyond a lot of people's reach), but it is not reasonable for others to use the plane to skip off for a weekend's sunning on the beach or to go to an exhibition or a concert. However, this is just what has been happening and it now seems that the notion of what is reasonable and what is not has become extremely blurred.

It is, however, going to be vital for us to give close thought to the subject if we want to be able to travel in a way that is compatible with our fossil-fuel reserves and which is also compatible with an eventual slowdown in the climate change we are currently experiencing.

As regards transport, this comes down to three main points:

• avoiding unnecessary travel and long-distance transport of consumer products and food
• making maximum use of public transport such as trains and trams, which are not heavy consumers of fossil fuel
• using low-consumption vehicles and, if possible, non-hydrocarbon fuels for transport considered as necessary.

The last point is possibly the most important and most misunderstood. The idea is not to rush around trying to find an immediate replacement for oil, as oil-based fuel has the physico-chemical properties that make it ideal for our transport needs. What we do want to do is to make our reserves last as long as possible so we can enjoy sustained use of petrol and diesel for driving and kerosene for flying. And once supplies run out, we will have to be able to produce substitute products using methods that are not harmful to the envi-

Make maximum use of public transport such as trains and trams, that are low consumers of fossil fuels.

After vehicle fuel consumption has been brought down, by reducing car weight, size, speed, acceleration and power, the future will probably feature the widespread adoption of hybrid vehicles that can charge up directly by plugging into the local electricity network.

ronment or the climate. These methods will use heat and hydrogen which we will have to produce using renewable and nuclear energy.

The solution is not, contrary to what most people and indeed a lot of large companies (such as Mercedes, BMW, General Motors, Renault, Shell, BP, Air Liquide and Linde) think, to rely on hydrogen for our long term needs for powering our cars. The long-term scenario, after vehicle fuel consumption has been brought down (by reducing car weight, size, speed, acceleration and power) will probably feature the widespread adoption of hybrid vehicles. Most of these vehicles are likely to be diesel-electric hybrids that can charge up directly by plugging into the local electricity network.

The widespread use of these vehicles will help solve the problem of energy resources and will also contribute to dampening the effects of climate change by spearheading the switch in our energy consumption away from fossil fuels to electricity produced by other sources.

As can be seen, and contrary to the conventional wisdom which claims that nuclear power can never solve the issue of our transport problems, the notion of "reasonable and sustainable mobility" is entirely compatible with a future based on nuclear energy.

Our aim here is not to draw up a list of the likely effects of climate change, which were set out clearly and simply in Al Gore's celebrated documentary film «The Inconvenient Truth», which was released in late 2006.

How can we enjoy a relatively stable climate

Climate change is probably the biggest risk currently facing mankind.

The risk of a sharp increase in average temperatures has been analyzed and confirmed many times over by the global scientific community. Most estimations put the rise at about 4°C by the end of the 21st century, which is the same difference, but in the opposite direction, as that experienced between the present day and the last great ice ages. It is quite possible that the effects of these changes could be worse than we imagine today. A lot of the changes are self-accelerating (scientists often call them positive feedbacks), somewhat similar to the way nuclear chain reactions work. The melting of snow and glaciers, for example, reduces the amount of solar radiation bounced back into space and hence helps increase global warming. The melting of the Siberian permafrost is causing methane to be released into the atmosphere, which also contributes to warming. Global warming itself encourages the production of methane in humid agricultural zones and reduces the solubility of carbon dioxide in the oceans. These are just a few of the many examples of self-accelerating phenomena related to climate change and global warming.

The predictions of a catastrophic melting of the world's surface ice-sheets over a relatively short period of time have now become plausible. The effect of the disappearance of the ice-sheets on the rise in the levels of the oceans can be measured quite accurately and comes out at about 80 meters, as shown in the map opposite. You certainly do not have to be a specialist to understand that such a rise would mean the end of the world as we know it and we could say good-bye to New York, London, Paris, Rome, Cairo and Shanghai, amongst others. Enormous population shifts and epidemics would be followed by a reduction in the world's population of around 75 – 90%. It is even possible that humanity might disappear, ravaged by disease or destroyed by fighting for the remaining areas of arable land.

Some 65 million years ago, dinosaurs disappeared in the midst of a vast period of natural climate change that saw the transition from the cretaceous to the tertiary period. The 21st century could equally well see the end of our own species, but this time the climate change, far from being a natural phenomenon, would be entirely our own fault.

Before a catastrophe of this scale can actually occur, climate change will already start to have a significant although gradual effect, and we have already begun to see this in our flora and fauna. Other consequences will have a more immediate impact, such as changes in rain cycles, with a higher frequency of flooding and drought, and changes in atmospheric pressure leading to a greater frequency of tornadoes and cyclones.

The prospect of these major upheavals should make us take a long, hard look at the risks we are taking in disturbing the fragile balance that allows life to exist on Earth. Our planet is the only one in our solar system that contains oxygen in its atmosphere and this oxygen has accumulated over time as a result of the development of life and through photosynthesis. Life and biodiversity depend on this atmosphere, which also acts as a filter for the solar radiation that we receive and which regulates the surface temperature of the Earth through the greenhouse effect. The atmospheric layer is, however, very thin, of very low mass and very fragile. If Earth were an apple, the thickness of its atmosphere would be the same as that of the skin covering it.

You can get a good idea of the mass of the atmosphere if you think about diving. When you dive down to 10 meters, the pressure is 1 bar, or one atmosphere*, higher than on the surface. This means that if the atmosphere were to be squeezed around the surface of the Earth, its mass would be no greater than that of a strip of water 10 meters deep.

Rise in sea-levels due to the rapid melting of Earth's ice-sheets.

Antarctic ice-sheets:	73m
Greenland ice-sheets:	6m
Other terrestrial ice-fields (Iceland, Spitzberg and glaciers):	1m
Total :	80M

Sources/ USGS (US Geological Survey).

The map below shows what will happen if the oceans rise 80 meters compared to their current depths.

*To be more precise, one atmosphere = 1013 millibars.

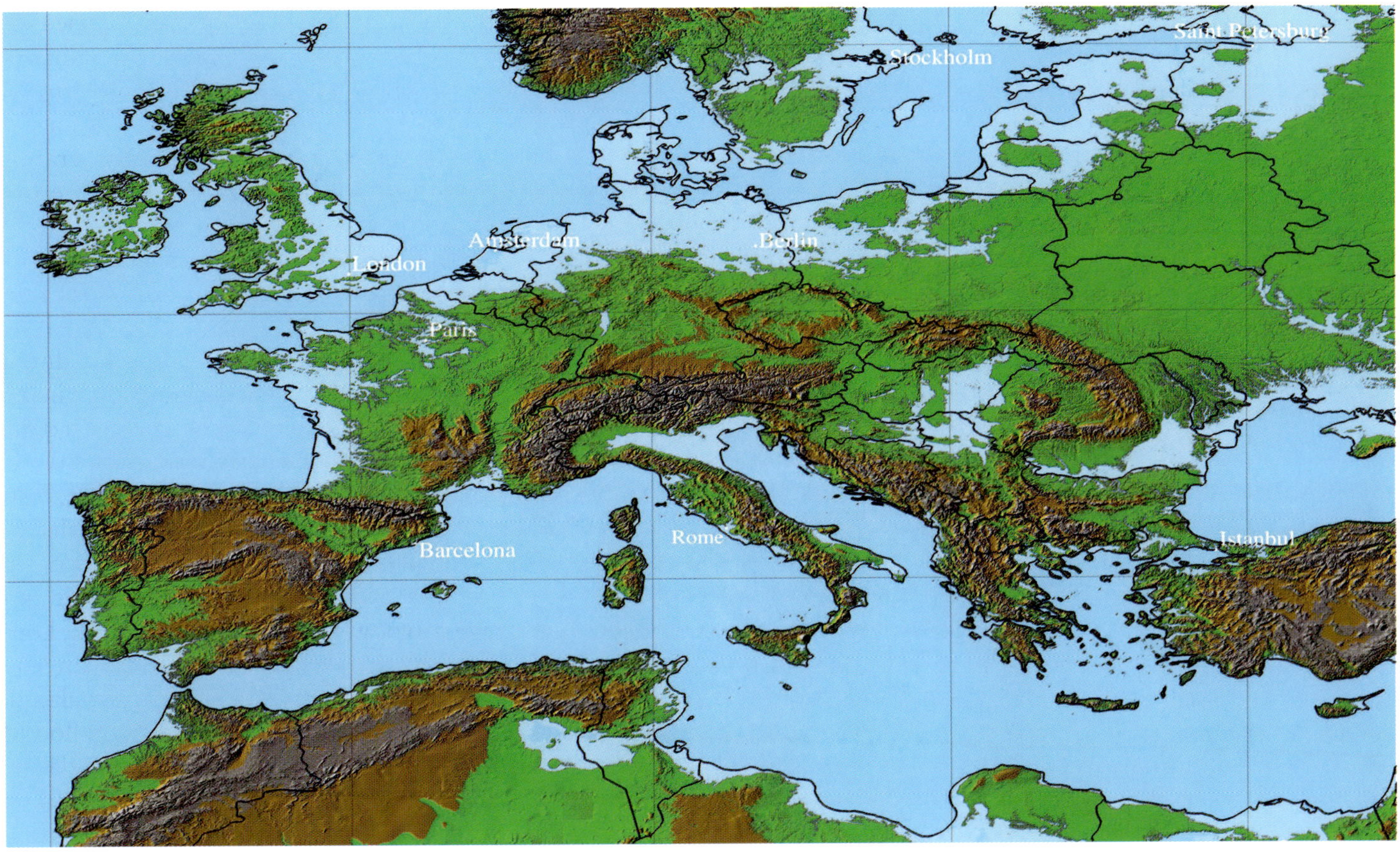

Hurricane Katrina building up off the coast of Louisiana on August 28, 2005.

As we can see, if humanity wants a sustainable future, not to mention sustainable development, it has no choice but to bring carbon dioxide emissions down both quickly and substantially. CO_2 must be captured and stored as much as possible, but above all, we have to reduce our consumption of hydrocarbon energies.

Apart from cutting energy consumption, there are only two ways open to us for reducing our dependence on fossil fuels: renewable energy and nuclear power. Given the scale of the problem, nuclear power is the only solution that can realistically be considered.

Renewable energies certainly represent an alternative, but it is very difficult to see how in the next 50 to 100 years we are going to use them to generate energy, in particular electricity, in the quantities that will be required. It is, nonetheless, essential to continue work in this field, even if there is no solution for the moment.

If humanity wants a sustainable future, not to mention sustainable development, it has no choice but to bring carbon dioxide emissions down both quickly and substantially

How can we have a reasonable life expectancy?

It would not be feasible to continue making progress in increasing our life expectancy if we are unable to come up with answers to these needs:

- ensuring an adequate food supply
- ensuring an adequate water supply
- reasonable possibilities of travel
- a healthy environment
- a relatively stable climate

There is, however, no use even thinking about the solutions if we are unable to meet our future energy requirements.

We have a seen a spectacular improvement in life expectancies in the richer countries over the last hundred or hundred and fifty years. When we try to analyze the reasons behind it, we immediately think of the progress that has been made in hygiene and in medicine, but it is not often that energy comes to mind. Energy has been, however, a major driver in increased life expectancy: it has improved working conditions and taken a lot of the drudgery and physical effort out of our jobs, it has helped drastically reduce child labor and it has also played a major role in improving our non-professional life.

Energy is, of course, essential for providing heat, but its effect is more far-reaching than is immediately obvious. Our ancestors used to use wood for heating and the poor ventilation was often an open invitation for respiratory disease and cancer. The ready availability of hot water (baths, showers, washing hands, etc.) has made a big contribution to better hygiene and this again is dependant on having a steady supply of energy. Sanitary conditions have been greatly helped by replacing oil-lamps and candles first by gas lighting and then by electricity. Food quality, diversity and conservation has also seen a significant change for the better with the introduction of domestic refrigerators and the commercial cold chain and this has had a noticeable effect on our general health.

These groundbreaking changes, all of which were made possible by modern energy sources, have extended our life expectancies. The use of air-conditioning is also having an effect, especially in the wake of the heatwaves of recent memory.

Modern healthcare systems also owe a lot of their efficiency to a stable supply of efficient energy. The fast transport of patients and doctors, the lighting in the operating theatres, electrosurgical instruments, lasers, medical imagery and incubators for prematurely-born or fragile babies are just a few of the enormous variety of energy-dependent developments we have seen over the past few decades.

The question that now needs to be answered, though, is if it is possible to continue this spectacular progress as the price of energy soars at the same time as supplies begin to dwindle. Even if our objective is simply to maintain a "reasonable" life expectancy, as opposed to increasing it, the answer will need to come fairly quickly.

The question also needs looking at from several different angles. If (and this is a hypothetical example) the price to pay to provide the extra energy needed to extend life expectancy by say one year is an acceleration in climate change, an increase in cyclones and flooding, it is not sure that such a solution would find many takers, especially when we consider that what is gained in life expectancy on the one hand could easily be lost a few years later as a result of climatic deterioration.

There's more to it than just electricity

As we have seen, we can only find sustainable solutions to the great challenges facing mankind if we can be sure of a sufficient supply of energy. We have also stressed that these energies must come from either renewable or nuclear sources.

Nuclear energies actually offer a fairly wide range of possibilities, although we do not as yet know what their development will lead to in the future. Three main options exist and these are breeder fission reactors, non-breeder fission reactors and possibly fusion reactors. All three have to be failsafe and offer a very high degree of safety and security.

We urgently need to put considerable resources into extending our knowledge of these three families, each of which, as we have seen throughout this book, can have a number of variants.

It would also be meaningless to pit renewable energies against nuclear energy. Excluding traditional biomass (mainly major hydroelectric projects), renewable energies account for just over 7% of worldwide energy supply in commercial primary energies and the figure is about the same for nuclear power. The total, then, amounts to about 15%, while the

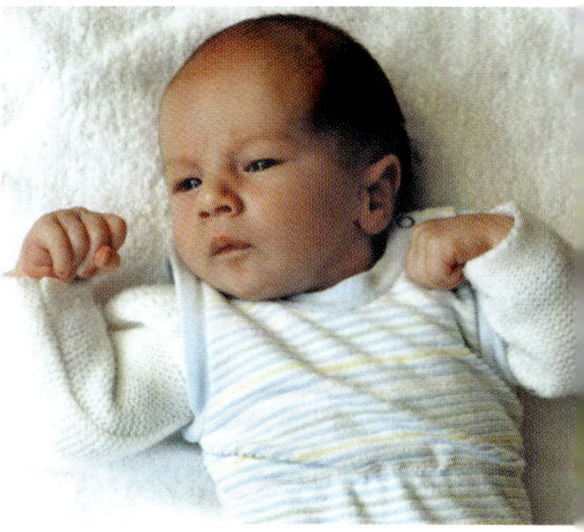

Energy is also central to the efficiency of modern healthcare systems, from medical imagery to incubators for prematurely-born or fragile babies

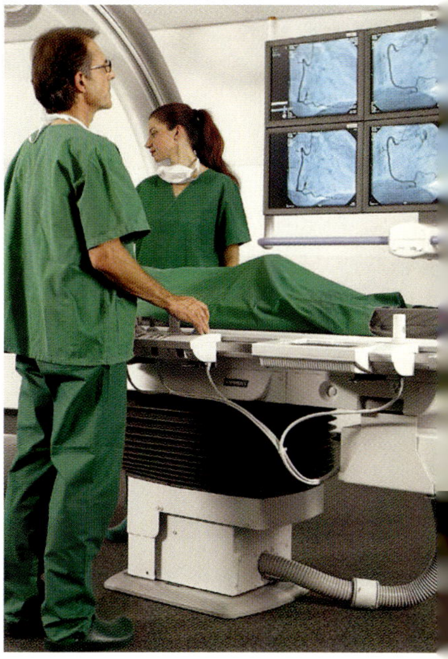

Waste from the Al-Doha seawater desalination plant in the Jahra region of Kuwait (opposite).

After long having to depend on Iraq for its water supply, desalination plants now cover 75% of Kuwait's needs. After flash distillation, water unfit for consumption is pumped out into the sea, where, like a giant octopus, it flows into the Persian Gulf. There are some 12,500 desalination plants in the world, spread over 120 countries and supplying about 20 million m³ of fresh water every day, or about 1% of the water consumed worldwide. Two thirds of this desalinated water comes from seawater and the rest from brine. Half of the plants are to be found around the Persian Gulf, in oil-producing countries which have the resources to supply these highly-expensive, energy-hungry factories. About one metric ton of fuel is burnt for every 100 m³ of fresh water produced. With water now becoming a worldwide issue of increasing concern, water consumption in Europe currently stands at about 250 to 350 liters per head per day (simply flushing a toilet uses about 6 liters of water).

fossil fuels these two energies are being called upon to replace represent about 85% of our current commercial primary energy consumption.

Renewable and nuclear power are therefore not in competition with each other but are fully complementary. Each has to find its own market to which it can offer appropriate, sustainable solutions that will have as little effect as possible on the environment. It should be borne in mind, however, that the production of electricity is not the only issue that needs addressing, especially when one considers that electricity currently only provides about a sixth of worldwide final energy consumption (1 Gtoe).

Heat requirements for industry, housing, offices, schools and hospitals are three times higher, at 3 Gtoe, than electricity needs, while transport accounts for another 2 Gtoe. Non-electricity needs represent some five-sixths of the world's final energy consumption and there can be no sustainable development if renewable and nuclear energies are unable to supply a large part these of these needs. Meeting a significant proportion of our heat and mobility energy needs are the among the future applications of nuclear power..

Future Applications of Nuclear Energy

Nuclear energy in the 21st century will first of all aim to meet our electricity needs. Electricity generation is currently by the far the most important role of nuclear power as an energy source, while the other energy applications we mentioned before, such as submarine and ship propulsion, take a back seat.

Future needs for nuclear-generated electricity are often underestimated. Twenty or thirty years from now, when the need to reduce our energy consumption begins to seriously bite as fossil resources begin to fall off (oil and gas peaks) and climate change continues to threaten, there will be a fast-track switch to electricity as the answer to our energy needs.

To answer these twin challenges, fossil fuels can of course be dismissed out of hand, although coal could still play a role if a way is found by then to capture and sequestrate CO_2 on a large scale. The only remaining solution to meet the massive increase in demand for electricity will once again be renewable and nuclear energy sources. New applications will in all likelihood be found for nuclear energy, outside the electricity sector. The high temperatures produced by the new generation of nuclear power plants will make them perfectly suited to a number of industrial applications (see diagram on the opposite page). Three examples of new uses of nuclear energy are given below:

- production of fresh water (desalination)
- heat production (for industry and housing)
- hydrogen production (for refineries, the petrochemical industry, the production of synthetic fuels and as an energy vector).

We are now going to go through these three potential new ways of harnessing nuclear energy.

This table gives our projection of the electric and non-electricity uses of nuclear energy over the 21st century

The Future of Nuclear Power	2005	2050	2100
Electrical uses in %	100 %	90 %	70 %
Thermal uses	-	3 %	10 %
Water desalination	-	3 %	10 %
Hydrogen production	-	3 %	10 %

Nuclear Energy and the Production of Fresh Water

Worldwide use of catching and canalized fresh water comprises 70% for agriculture, 20% for industry and 10% for domestic applications (of which less than 1% is for drinking and cooking).

The importance of a proper supply of

freshwater and especially drinking water is obvious and water, along with energy and food, is the key as to whether a population is developed or under-developed.

This can be easily illustrated with a couple of figures:

- in 2005, some 20% of the planet's population had no access to drinking water (estimates put the number of deaths from drinking contaminated water at 3 to 4 million per year).
- another 20% do not have enough fresh water to cover their daily needs.

These figures are unfortunately set to get worse over the next few decades, especially for areas where there is a lack of fresh water resources, unless their needs can be met by massive implementation of desalination plants.

There is nothing new about desalination, and history relates that Caesar's legionnaires used to evaporate fresh water from the water of the Mediterranean by using solar energy, possibly the first record of an ecological approach to solving the problem of a water shortage.

The world currently possesses about 12,500 desalination units, half of which are to be found in the Near or Middle East. The first desalination plant was built in Kuwait in the 60s. Global capacity is about 30 million m^3 per day, with the biggest plants (in Saudi Arabia and Israel) producing some 500,000 m^3 per day. Some countries have a total dependence on desalinated water, such as Saudi Arabia, Kuwait, the United Arab Emirates, Qatar and Israel, and also other countries such as Malta and the Canary Islands, which get 50% of their fresh water from desalination. Algiers will soon will soon also be supplied with fresh water from the desalination plants that are currently being built.

The most important obstacle facing a more generalized take-up of the desalination option is primarily one of cost, with high up-front and operating costs, especially given the large amounts of energy needed to power the plants.

Desalination plants operating in the world

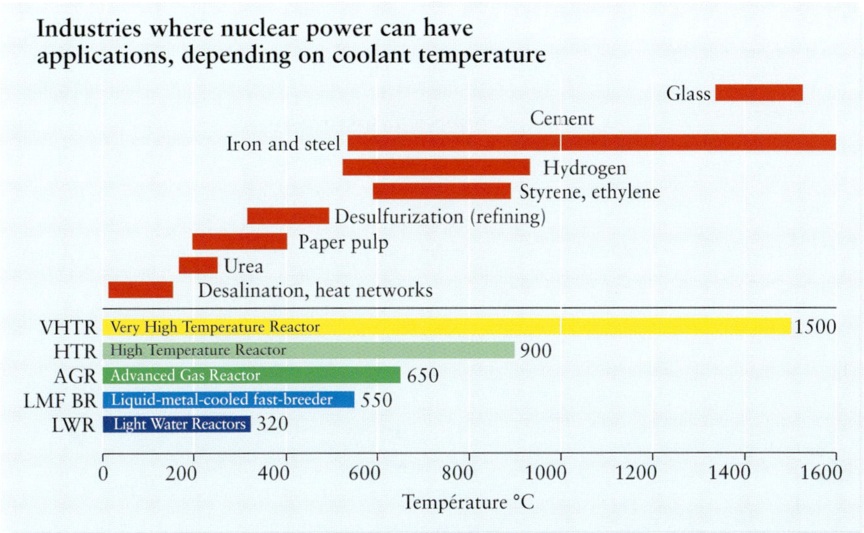

Yann Arthus-Bertrand: Earth from the Air.

Future needs for nuclear-generated electricity are often underestimated. Demand is also going to increase on the back of new applications such as the production of heat, fresh water and hydrogen

are practically all powered by fossil fuels or by using heat produced as a coproduct of electricity generating plants running also on fossil fuel (cogeneration).

The plants are generally to be found in the wealthier countries, as can be seen from the figures below:

- Middle East 60%
- USA and Mexico 15%

others (Spain, China, Japan, Australia, North Africa) 25%

There is nothing new about the idea of using nuclear energy for producing fresh water and considerable.

The production of fresh water through desalination might need up to 100 or 200 1,000 MW nuclear reactors by 2050 and up to 1,000 by the year 2100, in other words about 10% of the world's nuclear capacity by the end of the century.

Nuclear Energy and Heat Production

Bilibino in north-eastern Siberia is currently the only place where nuclear energy is being harnessed to produce heat for urban heating (see p.94).

With the limitations that are going to be placed on the use of fossil fuels and the rising price of such fuels due to the cost of CO_2 emissions in the atmosphere, it will soon be possible to use nuclear energy on a large scale to produce and market heat directly. The production could come in several forms and be aimed both at the domestic heating market and at certain industries which have "concentrated" needs :

- the oil industry: refining and production of ultra-heavy crudes such as tar sand, oil shales and production of hydrogen,
- the petrochemical industries: production of olefins (ethylene, styrene, vinyl),
- the gas and synthetic fuels industry: heat for the GTL (Gas to Liquids) and BTL (Biomass to Liquids) processes for producing synthetic fuels,
- cement works and the ceramics and glass industries,
- metallurgical industries: steel and non-ferrous metals,
- the paper industries.

It is not currently possible to put a figure on the uses of nuclear heat for urban networks and industry, but they could also account for up to about 10% of the worldwide installed nuclear capacity by the end of the 21st century, with fifty or so 1,000 MW plants in operation by 2050 and perhaps a few hundred over the fifty following years.

Desalination using nuclear power: Acquired experience

- In Kazakhstan, the Aktau BN-350 nuclear reactor operated for 27 years (from 1973 to 1999) and produced some 80,000 m^3 of potable water per day.
- In Japan, there are 10 desalination units linked up to PWR nuclear reactors (these are small units producing 2,000 to 3,000 m^3 /day of fresh water).
- In India, the 170 MW Kalpakkam nuclear reactor has been driving a desalination unit since the 1970s.
- In South Korea, a desalination project for producing 40,000 m^3/day, powered by a 330 MW PWR has reached an advanced stage.
- Other projects are in the pipeline in the Russian Federation, Pakistan, China and Morocco.

Hydrogen Production

Hydrogen will in all probability represent the third major development path (with heat production and the desalination of seawater) for nuclear energy on top of its traditional role in electricity production. Nuclear power can be used to produce hydrogen in two ways:

- through electrolysis
- through high-temperature thermochemical cycles where the reactant is water and the net products hydrogen and oxygen.

We will not be going into any detailed description of these methods for producing hydrogen from water, nor will we be making any comparisons of their technical and economic advantages. Suffice it to say that at the beginning of the 20th century, industrial production of hydrogen was carried out through electrolysis, to be used in the production of ammonia for nitrogen-based fertilisers (this is how the Norwegian group Norsk Hydro started up). This used to be the most economical way of producing hydrogen, and once the fossil-fuel method of producing the gas using hydrocarbons (by reforming natural gas or through the partial oxidation of oil residues) becomes more costly than producing it from electricity or nuclear heat, it will come back into widespread use. It is difficult to say with any great precision when the hydrocarbon-fuelled method of producing hydrogen will become more expensive than the alternative of making it from water and nuclear heat, as there are a number of factors in play, such as the future price of hydrocarbons and also the cost of CO_2 emissions. Without going too far into the calculations, we can say that hydrogen production from nuclear energy should become economically viable once the oil peak has been passed, in other words around 2015 - 2030.

This new market for nuclear power might well prove to be considerable, but we should not let ourselves get carried away by the optimism of people who think that hydrogen will be fuelling the cars of the future. Even if we manage to sort out all the safety issues relating to the distribution of hydrogen and its use in hundreds of millions of vehicles, we think that economic factors will prevent it becoming used on any large scale.

In our opinion, the main use of hydrogen in the long term will be in producing synthetic hydrocarbons, as we already mentioned before. Before that, however, hydrogen produced using nuclear energy will probably, oddly enough, find its primary market in the oil industry, especially as demand for hydrogen in the sector is steadily rising with the need to refine increasingly heavy crudes before converting them into lighter, cleaner products.

Hydrogen appears to be heading towards the role of a "technological bridge" between the nuclear and oil industries.

It is again difficult to put a figure to the size that this new application of nuclear energy might reach. We can only hypothesize that this market should, by the end of the century, account for about the same proportion of demand for nuclear energy as the production of heat and the desalination of sea water.

Mercedes brought out its first fuel-cell-powered vehicle, the NECAR, 1 in 1994 and 5 years later they announced that the technology would be available for their customers. The first big deliveries were made in Berlin in 2004.

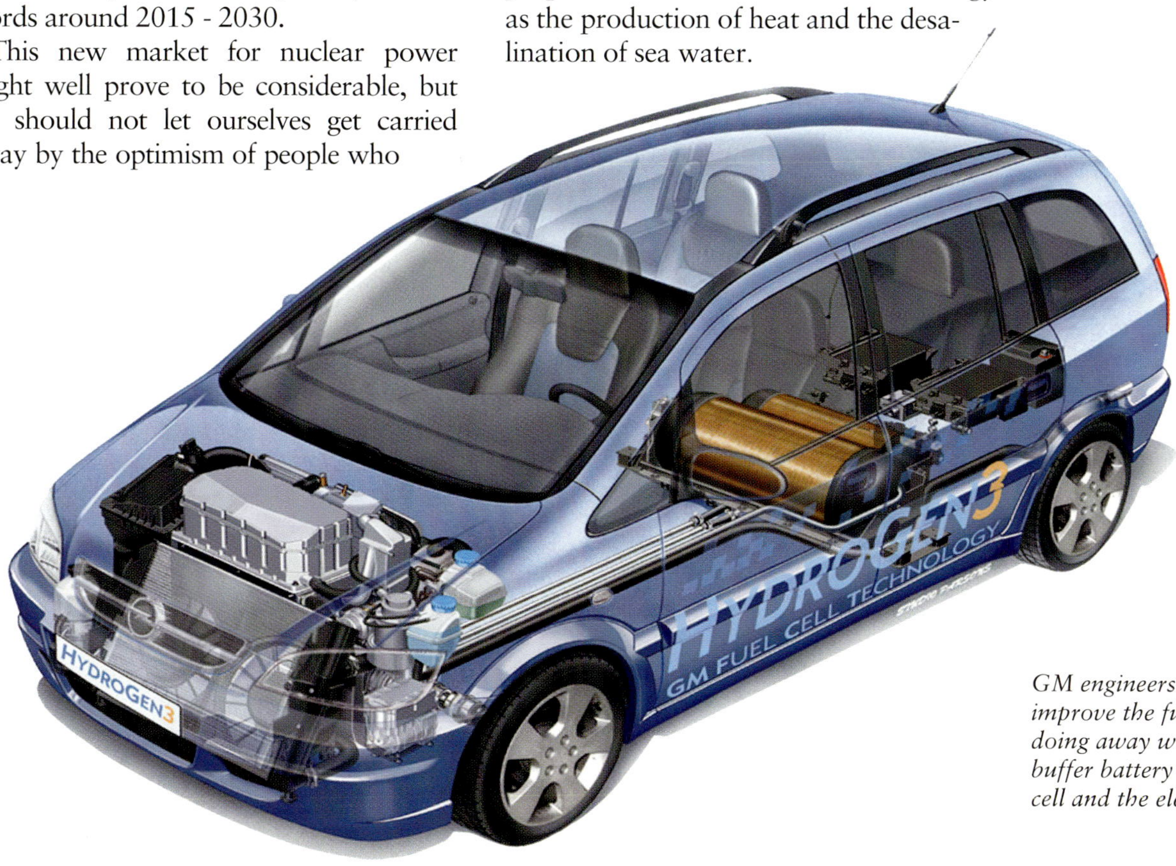

GM engineers have managed to improve the fuel-cell system by doing away with the need for a buffer battery between the fuel cell and the electric motor.

Conclusion

Over the coming years, mankind is going to have to face up to two major issues that are of paramount importance for its future. The first is the inevitable decline in worldwide production of oil and natural gas and the second is climate change.

The 21st century will see considerable building of technological bridges between the various energy sectors

There can be no solution to these two problems if the world does not act in unison. An "each for his own" policy would without doubt lead countries into a violent race to gain control of the remaining oil and gas resources, while the lack of a concerted energy policy would take us straight into a climate disaster of cataclysmic proportions. Where energy issues are concerned, the 21st century therefore needs genuine governance on a world scale, unless humanity wants to find itself on a lemming-type march towards a vast collective suicide.

Trusting that common sense will prevail, the first step in facing up to these two fundamental problems will be adopting a much more reasonable approach to energy consumption and making increasing use of renewable energy sources. This, of course, will not be enough in itself and we will also have to turn to the massive use of nuclear energy. As the figures we gave before illustrate, this is a huge task which is going to require immense concerted efforts with regard to research into the next generation of reactors and the corresponding nuclear fuel cycles, to the reduction in waste products and their safe storage and to the prevention of proliferation of nuclear weapons.

This mammoth endeavour cannot just be left up to individual governments or to market forces. The enormous development of nuclear energy will require massive investment, with powerful industrial operators acting within a worldwide market. The development will also need vast research programs as well as regulatory control involving safety and security issues, and this a role which governments will have to fulfil.

As with every project on this kind of scale, there are inherent risks involved and there is even the possibility of serious accidents. We will, however, have to strive to accept these risks and take every possible precaution to prevent them from turning into reality. Sustainable development, or put more starkly, our survival, has its price.

No human action is without risk, but the risk involved in taking no action at all is a lot more threatening for our future.

The 21st century will see considerable building of technological bridges between the various energy sectors. Renewable ener-

Symbolic of the future, these three pictures show how energy could be revolutionized in the 21st century, with the complementary interaction of fossil, renewable and nuclear energies. The century will also see increased electrification to meet our energy needs.

gies from solar and wind technologies will need continuous sources of non-CO_2-emitting energy to make up for variations and shortfalls in their power generation and they will have to work closely with the nuclear industry. Fossil fuels will have to follow the same path as they will need also huge amounts of non-CO_2-emitting energy, both at front-end (improving recovery rates) and at back-end (refining). The oil industry will also be demanding nuclear energy for producing the hydrogen required for refining heavy and conventional crudes, for its petrochemical activities and for producing synthetic fuels.

The large-scale use of biomasses will also in all likelihood need the support of nuclear power. Yields per hectare of crops grown for producing biofuels could double or even triple if they use the heat and hydrogen produced by nuclear energy.

This vision of the 21st century, based on a sensible approach to consumption, concerted action at international level, the massive use of nuclear energy and its combination with other energy sources, is of course only the authors' vision.

We are, however, convinced that this type of future is possible and it is a future that allows real hope for the sustainable development of human societies.

Photo Credits

Agence Corbis
Agence Fotolia
Agenge Gamma
Agence Getty
Agence Magnum
AIE/IEA
AIEA/IAEA
AFP
Aida Nori
Andra
AREVA
Artus Bertrand Yan
Automobiles Peugeot
Bannister Barbara/Corbis
Barré Bertrand
Benali Remi
Bettmann/Corbis
Brack Dennis
British Columbia,
Salmon Farmers Ass.
Carnemolla John/Corbis
Cape Kennedy
CDU
CEA
City of Hong Kong
Cycles Peugeot
Daimler Benz
Demange Francis
D.Lepp George/Corbis
Diordiev Constantin
Durand Patrick/Corbis
EDF
EPIA
ESLAMI RAD/GAMMA
Eurodif
Fobes Natalie/Corbis
Franken Owen/Corbis
Getty images
General Electric
General Motors
Gonin Antoine
Gore Al
Gruyaert H/Magnum
Hirlé Lionel
Hirlé Ronald
Hoffman Heinrich
Hoepker Thomas/Magnum
Jannes Y./ AREVA
Kaysersberg cartonneries
Kostin Igor/Corbis
Lauten Daniel Sannum
Lawrence Laboratory
Lefkowitz Lester/Corbis
Lemmens Frans/Corbis
Lesage P./ AREVA
Los Alamos
Lovelock James
Mc Pherson Colin
Museum of Hiroshima
Museum of Nagasaki
Nasa Visible Earth
O'Rear Charles/Corbis
NATO
Palix O./ AREVA
Ressmeyer Roger/Corbis
Roosevelt Library
Rotkin Charles/Corbis
Reuter/Corbis
Sandia Laboratory
SGL Brakes GmbH
Starostenko Sergey
Senator Domenici
Site of Handford & d'Oakridge
Site of Olkiuoto
Siemens medical
Smirnov Vladimir/Corbis
Sodel/EDF
Sohm Joseph/Corbis
Stracel
Taillat Jean-Marie
Three Mile Island, nuclear
power plant
Time magazine
Truman Library
University of Wisconsin
University of Georgia
Usine G. Besse (Tricastin)
US Army archives
U.S. Navy
U.S. Air Force
Vanderwerff Steven H.
Wright W./ AREVA
Wright Tim/Corbis
Yamal
Zaklin Stefan
Zajic Miroslav/Corbis
Photos X

Illustrations :

Ashcroft C. Mark
Gen IV
International Forum
Klotz Jean-François
Nast Philippe

Acronyms and abbreviations

AGR	Advanced Gas Reactor, gas cooled reactor of British design.
ALARA	As low as reasonably achievable
BNFL	British Nuclear Fuel Ltd
BWR	Boiling Water Reactor.
CANDU	Heavy water reactor (Canada Deuterium Uranium)
CEA	Commissariat à l'Energie Atomique (France)
DC	Developing Countries
DOE	US Department of Energy
EDF	Electricité de France, the largest nuclear plant operator
EPR	Evolutionary Pressurized water Reactor, 3rd generation PWR of AREVA design.
EU	European Union
FBR	Fast Neutron Breeder Reactor
GE	General Electric (BWR designer)
GFR	Gas-cooled FBR
GHG	Greenhouse Effect Gas (CO_2, CH_4, O_3, N_2O, etc.)
HLW	High-level radioactive waste.
HTR	High-Temperature Reactor
IAEA	International Atomic Energy Agency.
ILW	Intermediate level radioactive waste
IPCC	Intergovernmental Panel on Climate Change.
LFR	Lead-alloy-cooled FBR
LL-ILW	Long-lived ILW
LLW	Low-level radioactive waste
LWR	Light Water Reactor (PWR, VVR and BWR)
MAGNOX	Natural Uranium Gas-cooled reactor of British design
MOX	Plutonium bearing fuel for LWR (Mixed Oxide)
MSR	Molten-Salt Reactor
NNWS	Non-Nuclear-Weapon States: NPT parties outside the 5 NWS
NPP	Nuclear Power Plant
NPT	Non-Proliferation Treaty
NWS	Nuclear-Weapon States (USA, Russia, UK, France and China)
OECD	Organisation for Economic & Commercial Development
PWR	Pressurized Water Reactor
R&D	Research & Development
RBMK	Graphite Boiling Water Reactor of Russian design
RRA	Reasonably Assured Resources
SCWR	Reactor cooled by supercritical ordinary water.
SFR	Sodium-cooled FBR
SWU	Separative Work Unit
TBP	Tri-butyl Phosphate, solvent used in uranium chemistry
toe	Metric ton of Oil Equivalent
UNGG	French designed MAGNOX (Uranium Naturel Graphite Gaz)
VHTR	Very High Temperature Reactor
VVR	Russian-designed PWR

Glossary of terms

ACTINIDE
(See Transuranic element)
Chemical element with a nucleus containing more than 88 protons. In order, the actinides are actinium, thorium, protactinium, uranium and the transuranic elements. Neptunium, americium and curium are often referred to as minor actinides.

ASSEMBLY, FUEL ASSEMBLY
(See Fuel element)

ATOM
Basic component of the chemical elements that form matter. It consists of a nucleus composed of positively charged or neutral particles (protons and neutrons), orbited by negatively charged particles (electrons).

CENTRIFUGE PROCESS
(OR ULTRACENTRIFUGATION)
Enrichment process in which a gaseous mixture of isotopes is spun at very high speed, using the centrifugal force to modify the composition of the mixture.

CLADDING
Leaktight metallic tube which surrounds the nuclear fuel to protect it against water corrosion and prevent the migration of the fission products.

CONDITIONING
Waste conditioning: operation whereby waste is converted into a form suitable for transportation and/or storage and/or final disposal.

CONTAMINATION
(See "Radioactivity")
Presence of an undesirable level of radioactive substances (dust or liquid) at the surface of or inside any medium. Contamination in humans can be external (on the skin) or internal (via the respiratory or digestive tracts).

CRITICALITY
A medium containing a fissile nuclear material becomes critical when neutrons are produced (by the fissions of this material) at the same rate as they disappear (through absorption and leakage to the outside).

DECOMMISSIONING
Term covering all the steps following the shutdown of a nuclear or mining facility at the end of its operating life, from closure to the removal of radioactivity on the site, and including physical dismantling and clean-up of all non reusable facilities and equipment.

DOSE
Measurement characterizing the exposure of individuals submitted to radiation.
• Absorbed dose: quantity of energy absorbed by matter (living or inert) exposed to radiation. It is expressed in grays (Gy).
• Dose equivalent: in living organisms, an absorbed dose has different effects depending on the type of radiation (X, alpha, beta and gamma). To take these differences into account, a dose multiplying factor is used (known as the "quality factor") to compute a "dose equivalent".
• Effective dose: sum of weighted dose equivalents deposited on the different tissues and organs by internal and external irradiation. The unit of measure for effective dose is the sievert (Sv).
• Lethal dose: fatal dose of nuclear or chemical origin.
• Maximum permissible dose: dose that must not be exceeded for a given period of time.

ENRICHED AND DEPLETED URANIUM
Before it is used to manufacture fuel elements, natural uranium is enriched with U235 (the proportion of U235 is then 3% to 5%) Uranium enriched in U235 is obtained from natural uranium using an isotope separation process. The physical or chemical processes used to produce enriched uranium also produce at the same time uranium that has a lower proportion of U235 than natural uranium: this is known as depleted uranium.

ENRICHMENT
Process used to increase the abundance of fissile isotopes in an element.

EURATOM
European Atomic Energy Community. Treaty signed in Rome on March 25, 1957, at the same time as the founding treaty of the EEC, creating the European Atomic Energy Community with a view to establishing "the conditions required for the formation and rapid growth of nuclear industries". At present it is composed of all 27 EU member states.

EXPOSURE
Exposure of an organism to a source of radiation characterized by the dose received.
• External exposure: exposure from a radiation source located outside the organism.
• Internal exposure: exposure from a radiation source located inside the organism.

FISSION
Splitting of a heavy nucleus – generally upon impact with a neutron – into two smaller nuclei (fission products), accompanied by the emission of neutrons and radiation*, and the release of a considerable amount of heat. The energy thus released as heat is the source of nuclear power.

FISSION PRODUCTS
Fragments of heavy nuclei produced by nuclear fission* (splitting of ^{235}U or ^{239}Pu nuclei) or subsequent radioactive decay of nuclides formed during this process. All fission fragments and their daughter products are called "fission products".

FUEL CYCLE
All the industrial operations undergone by nuclear fuel. These operations include extraction, processing uranium ore, conversion, uranium enrichment, fuel manufacturing, reprocessing spent fuels, and waste management. The fuel cycle is "closed" if it includes the reprocessing of spent fuel and recycling of fissile materials resulting from reprocessing.

FUEL ELEMENT (or assembly)
Assembly of rods joined together and filled with uranium or MOX pellets. Depending on the type of plant, the reactor core contains from 100 to 200 fuel assemblies.

FUEL ROD
Metal tube (about 4 m in length and 1 cm in diameter) filled with pellets (about 300) of nuclear fuel.

GASEOUS DIFFUSION (See "ENRICHMENT")
Process for separating molecular species, based on the difference in the velocity of these molecules (due to their different mass and dimensions) as they pass through a semi-permeable membrane.

ISOTOPES
Elements whose atoms* have the same number of electrons and protons but a different number of neutrons. A given chemical element can therefore have several isotopes with a differing number of neutrons. All the isotopes of a given element have the same chemical properties, but different physical properties (mass in particular).

MOX
("Mixed Oxides"): mixture of uranium* and plutonium* oxides used to make certain nuclear fuels*.

NATURAL URANIUM (Unat)
Naturally occurring radioactive element in the form of a hard, gray metal, found in several ores, pitchblende in particular. Natural uranium comes as a mixture composed of 99.28% fertile ^{238}U, and 0.71% ^{235}U.

NUCLEAR FUEL
Nuclide that releases energy when it is consumed by fission* inside a reactor*. By extension, any product containing fissile materials that supplies energy in a reactor core by sustaining the chain reaction.

NUCLEAR SAFETY
(See "SAFETY ANALYSIS REPORT")
All the measures taken at every stage of the design, construction, operation and final shutdown of a facility to ensure operational safety, prevent incidents, and limit their impact.

PLUTONIUM
Chemical element with the atomic number 94 and conventional symbol Pu. Plutonium-239, a fissile isotope*, is produced in nuclear reactors* from uranium-238.

RADIATION, IONIZING RADIATION
(See "Radioactivity")
Electromagnetic waves (such as radio waves, light waves, UV or X rays, cosmic rays, etc.), and particles (electrons, protons, and neutrons) or groups of these particles. The energy of these waves or particles is respectively proportional to their frequency and velocity. These types of radiation act on irradiated objects by stripping electrons from their atoms, leaving ionized atoms (electrically charged) in their wake, whence the generic name of IONIZING RADIATION.

RADIOACTIVE DECAY or DISINTEGRATION
Natural decrease in the nuclear activity of radioactive substance through spontaneous disintegration.

RADIOACTIVE HALF-LIFE
Time required for half the atoms* contained in a sample of radioactive substance to disintegrate naturally. The radioactivity* of the substance has therefore been halved. The half-life varies with the characteristics of each radionuclide:

RADIOACTIVE WASTE
Non-reusable radioactive by-products coming mostly from the nuclear industry, but also from laboratories, hospitals, etc.

RADIOACTIVITY
(See "Dose")
Emission by a chemical element of electromagnetic waves and/or particles caused by a change in its nucleus. Emission can be spontaneous (natural radioactivity of certain unstable atoms) or induced (artificial radioactivity). Radioactivity has several forms:
• Emission of "alpha* particles" (combination of 2 protons and 2 neutrons), called "alpha radiation".
• Emission of electrons, known as "beta* radiation".
• Emission of electromagnetic waves, called "gamma radiation".

RADIOLOGICAL PROTECTION
(See "Radioactivity")
All the measures intended to preserve the health of the general public and workers by protecting them against radiation and ensuring that legal provisions are observed.

RADIOELEMENT
Any radioactive substance. There are only a few naturally occurring radioelements, including some heavy elements (thorium, uranium, radium, etc.) and some light elements (carbon-14, potassium-40, etc.). The others, more than 1,500 in all, are created artificially in the laboratory for medical purposes or in nuclear reactors as fission products.

REACTOR, NUCLEAR REACTOR
Device in which controlled nuclear reactions are carried out. The heat released by these reactions is harnessed to form the water vapor used to operate a turbine driving an electric generator. Models vary according to the type of fuel, the moderator used to control the reaction, and the coolant used to remove the heat to be recovered.

REPROCESSING
Processing spent fuel to extract fissile and fertile materials (uranium and plutonium) for recycling purposes and to condition the different types of waste in a form suitable for disposal. Fission products and transuranic elements are vitrified.

SHIPPING CASK
Packing specially designed to ensure total containment of certain radioactive materials (spent fuel, vitrified waste, etc.) during their transportation, and to withstand any accidents.

TRANSURANIC ELEMENTS
(See Actinides)
Chemical elements in which the nucleus contains more protons than that of uranium. The first transuranic elements are, in increasing order, neptunium, plutonium, americium and curium.

URANIUM
Chemical element with the atomic number 92 and conventional symbol U, with two main isotopes: fertile 238U (99.3%) and fissile 235U (0.7%)

VITRIFICATION
Process used to solidify concentrated solutions of fission products and transuranic elements separated during spent fuel reprocessing by mixing them with a glass matrix at high temperature

YELLOWCAKE
Uranium concentrate (around 80%)

Web links for further information

General Information about energy

International Energy Agency AIE/IEA
L'AIE is a OECD Agency advising the governments of 26 member states in energy policy. IEA compiles and edits syntheses of the world energy data. IEA publishes regularly its "World Energy Outlook".
www.iea.org

World Energy Council CME/WEC
WEC has Member Committees in over 90 countries, including most of the largest energy-producing and energy-consuming countries. Established in 1923; the organisation covers all types of energy, including coal, oil, natural gas, nuclear, hydro, and renewables, and is UN-accredited, non-governmental, non-commercial and non-aligned. WEC's Mission is "To promote the sustainable supply and use of energy for the greatest benefit of all people"
www.worldenergy.org

World Wildlife Fund WWF
This large international environmentalist NGO is somehow representative of critics of nuclear power.
www.wwf.org

Energy Information Administration EIA
Official energy statistics from the US Government
www.iea.doe.gov

Eurostat
The statistical office of the European Communities
http://eurostat.ec.europa.eu

Information on climate change

Intergovernmental Panel on Climate Change (IPCC)
IPCC was established in 1988 under the auspices of the United Nations. Its role is to assess on a comprehensive, objective, open and transparent basis the scientific, technical and socio-economic information relevant to understanding the scientific basis of risk of human-induced climate change, its potential impacts and options for adaptation and mitigation.
www.ipcc.ch

Information on nuclear power (International)

International Atomic Energy Agency AIEA/IAEA. The IAEA is the world´s center of cooperation in the nuclear field. It was set up as the world´s "Atoms for Peace" organization in 1957 within the United Nations family. The Agency works with its Member States and multiple partners worldwide to promote safe, secure and peaceful nuclear technologies.
www.iaea.org/worldatom/

Nuclear Energy Agency AEN/NEA
The Nuclear Energy Agency (NEA) is a specialised agency within the Organisation for Economic Co-operation and Development (OECD). Its mission is to assist its Member countries in maintaining and further developing, through international co-operation, the scientific, technological and legal bases required for the safe, environmentally friendly and economical use of nuclear energy for peaceful purposes.
www.nea.fr/

European Nuclear Society ENS
ENS groups throughout Europe 26 learned societies dedicated to nuclear sciences and technology.
www.euronuclear.org

Nuclear Energy Institute NEI
Professional Association of American nuclear industries.
www.nei.org

World Nuclear Association WNA
Former " Uranium Institute ",
WNA is a Professional Association of nuclear industries.
www.world-nuclear.org

The World's Nuclear News Agency NUCNET
Press agency providing nuclear information over the internet
www.worldnuclear.org

AREVA
French industrial group with activities notably in Germany and the USA, AREVA is one of the world leaders in non-CO_2 energy sources.
www.areva.com

American Nuclear Society ANS
The American Nuclear Society is a not-for-profit, international, scientific and educational organization. ANS has a membership of 10,500 engineers, scientists, administrators, and educators. Its core purpose is to promote the awareness and understanding of the application of nuclear science and technology.
www.ans.org

Information on nuclear energy (France)

Société Française de l'Énergie Nucléaire SFEN
Both a learned society and an information agency on nuclear themes.
www.sfen.org

Environmentalists for nuclear energy EFN
EFN is in favour of all energies which respect the environment, including the peaceful applications of clean nuclear energy as well as renewable energies
www.ecolo.org

Commissariat à l'Énergie Atomique CEA
French nuclear R&D public Organization
www.cea.fr

Autorité de sûreté nucléaire ASN
Independent Authorithy in charge of the technical and regulatory control of nuclear Safety and Radiation Protection.
www.asn.fr

Information on renewable energy sources

Energy Efficiency and Renewable Energy
A Department of the US DOE
www.eere.energy.gov

RE-Energy
A site dedicated to teaching materials on renewable energies..
www.re-energy.ca